EGYPT

Bédouin de mont Sinai. Nº 130   Artistique G. Lekegian &

EXPLORATION OF AFRICA: THE EMERGING NATIONS

EGYPT

1880 TO THE PRESENT:
DESERT OF ENVY, WATER OF LIFE

DANIEL E. HARMON

INTRODUCTORY ESSAY BY
Dr. Richard E. Leakey
Chairman, Wildlife Clubs
of Kenya Association
✛
AFTERWORD BY
Deirdre Shields

CHELSEA HOUSE PUBLISHERS
Philadelphia
In association with Covos Day Books, South Africa

CHELSEA HOUSE PUBLISHERS

EDITOR IN CHIEF Sally Cheney
PRODUCTION MANAGER Kim Shinners
ART DIRECTOR Sara Davis
ASSOCIATE ART DIRECTOR Takeshi Takahashi
SERIES DESIGNER Keith Trego
COVER DESIGN Emiliano Begnardi

The Chelsea House World Wide Web address is http://www.chelseahouse.com
First Printing
1 3 5 7 9 8 6 4 2

Library of Congress Cataloging-in-Publication Data

Harmon, Daniel E.
    Egypt : 1880 to the present : desert of envy, water of life / Daniel E. Harmon; introductory essay by Richard E. Leakey ; afterword by Deirdre Shields.
        p. cm.—(Exploration of Africa, the emerging nations)
    Includes index.
    Contents: The dark continent / Richard E. Leakey—Foreigners conquer a once great nation—The canal—West meets East in the shadow of the pyramids—"Sharing" the government: the condominium era—A continent clamors for self-rule—A stormy independence—Egypt today—World without end / Deirdre Shields.
    Summary: Explores events in the recent history of Egypt, focusing on the political struggles that led from independence to foreign domination and back to independence.
    ISBN 0-7910-5744-5 (alk. paper)
    1. Egypt—History—9th century—Juvenile literature. 2. Egypt—History—20th century—Juvenile literature.
[1. Egypt—History—19th century. 2. Egypt—History—20th century.] I. Title. II. Series.

DT107 .H32 2001
962'.04—dc21   2001047239

The photographs in this book are from the Royal Geographical Society Picture Library. Most are being published for the first time.

The Royal Geographical Society Picture Library provides an unrivaled source of over half a million images of peoples and landscapes from around the globe. Photographs date from the 1840s onwards on a variety of subjects including the British Colonial Empire, deserts, exploration, indigenous peoples, landscapes, remote destinations, and travel.

Photography, beginning with the daguerreotype in 1839, is only marginally younger than the Society, which encouraged its explorers to use the new medium from its earliest days. From the remarkable mid-19th century black-and-white photographs to color transparencies of the late 20th century, the focus of the collection is not the generic stock shot but the portrayal of man's resilience, adaptability, and mobility in remote parts of the world.

In organizing this project, we have incurred many debts of gratitude. Our first, though, is to the professional staff of the Picture Library for their generous assistance, especially to Joanna Scadden, Picture Library Manager.

CONTENTS

EXPLORATION OF AFRICA: THE EMERGING NATIONS

THE DARK CONTINENT

DR. RICHARD E. LEAKEY

THE CONCEPT OF AFRICAN exploration has been greatly influenced by the hero status given to the European adventurers and missionaries who went off to Africa in the last century. Their travels and travails were certainly extraordinary and nobody can help but be impressed by the tremendous physical and intellectual courage that was so much a characteristic of people such as Livingstone, Stanley, Speke, and Baker, to name just a few. The challenges and rewards that Africa offered, both in terms of commerce and also "saved souls," inspired people to take incredible risks and endure personal suffering to a degree that was probably unique to the exploration of Africa.

I myself was fortunate enough to have had the opportunity to organize one or two minor expeditions to remote spots in Africa where there were no roads or airfields and marching with porters and/or camels was the best option at the time. I have also had the thrill of being with people untouched and often unmoved by contact with Western or other technologically based cultures, and these experiences remain for me amongst the most exciting and salutary of my life. With the contemporary revolution in technology, there will be few if any such opportunities again. Indeed I often find myself slightly saddened by the realization that were life ever discovered on another planet, exploration would doubtless be done by remote sensing and making full use of artificial, digital intelligence. At least it is unlikely to be in my lifetime and this is a relief!

Notwithstanding all of this, I believe that the age of exploration and discovery in Africa is far from over. The future offers incredible opportunities for new discoveries that will push back the frontiers of knowledge. This endeavor will of course not involve exotic and arduous journeys into malaria-infested tropical swamps, but it will certainly require dedication, team work, public support, and a conviction that the rewards to be gained will more than justify the efforts and investment.

Early Explorers

Many of us were raised and educated at school with the belief that Africa, the so-called Dark Continent, was actually discovered by early European travelers and explorers. The date of this "discovery" is difficult to establish, and anyway a distinction has always had to be drawn between northern Africa and the vast area south of the Sahara. The Romans certainly had information about the continent's interior as did others such as the Greeks. A diverse range of traders ventured down both the west coast and the east coast from at least the ninth century, and by the tenth century Islam had taken root in a number of new towns and settlements established by Persian and Arab interests along the eastern tropical shores. Trans-African trade was probably under way well before this time, perhaps partly stimulated by external interests.

Close to the beginning of the first millennium, early Christians were establishing the Coptic church in the ancient kingdom of Ethiopia and at other coastal settlements along Africa's northern Mediterranean coast. Along the west coast of Africa, European trade in gold, ivory, and people was well established by the sixteenth century. Several hundred years later, early in the 19th century, the systematic penetration and geographical exploration of Africa was undertaken by Europeans seeking geographical knowledge and territory and looking for opportunities not only for commerce but for the chance to spread the Gospel. The extraordinary narratives of some of the journeys of early European travelers and adventurers in Africa are a vivid reminder of just how recently Africa has become embroiled in the power struggles and vested interests of non-Africans.

THE DARK CONTINENT

AFRICA'S GIFT TO THE WORLD

My own preoccupation over the past thirty years has been to study human prehistory, and from this perspective it is very clear that Africa was never "discovered" in the sense in which so many people have been and, perhaps, still are being taught. Rather, it was Africans themselves who found that there was a world beyond their shores.

Prior to about two million years ago, the only humans or proto-humans in existence were confined to Africa; as yet, the remaining world had not been exposed to this strange mammalian species, which in time came to dominate the entire planet. It is no trivial matter to recognize the cultural implications that arise from this entirely different perspective of Africa and its relationship to the rest of humanity.

How many of the world's population grow up knowing that it was in fact African people who first moved and settled in southern Europe and Central Asia and migrated to the Far East? How many know that Africa's principal contribution to the world is in fact humanity itself? These concepts are quite different from the notion that Africa was only "discovered" in the past few hundred years and will surely change the commonly held idea that somehow Africa is a "laggard," late to come onto the world stage.

It could be argued that our early human forebears—the *Homo erectus* who moved out of Africa—have little or no bearing on the contemporary world and its problems. I disagree and believe that the often pejorative thoughts that are associated with the Dark Continent and dark skins, as well as with the general sense that Africans are somehow outside the mainstream of human achievement, would be entirely negated by the full acceptance of a universal African heritage for all of humanity. This, after all, is the truth that has now been firmly established by scientific inquiry.

The study of human origins and prehistory will surely continue to be important in a number of regions of Africa and this research must continue to rank high on the list of relevant ongoing exploration and discovery. There is still much to be learned about the early stages of human development, and the age of the "first humans"—the first bipedal apes—has not been firmly established. The current hypothesis is that prior to five million years ago there were no bipeds, and this would mean that humankind is only five million years old. Beyond Africa, there were no humans until just two million years ago, and this is a consideration that political leaders and people as a whole need to bear in mind.

RECENT HISTORY

When it comes to the relatively recent history of Africa's contemporary people, there is still considerable ignorance. The evidence suggests that there were major migrations of people within the continent during the past 5,000 years, and the impact of the introduction of domestic stock must have been quite considerable on the way of life of many of Africa's people. Early settlements and the beginnings of nation states are, as yet, poorly researched and recorded. Although archaeological studies have been undertaken in Africa for well over a hundred years, there remain more questions than answers.

One question of universal interest concerns the origin and inspiration for the civilization of early Egypt. The Nile has, of course, offered opportunities for contacts between the heart of Africa and the Mediterranean seacoast, but very little is known about human settlement and civilization in the upper reaches of the Blue and White Nile between 4,000 and 10,000 years ago. We do know that the present Sahara Desert is only about 10,000 years old; before this Central Africa was wetter and more fertile, and research findings have shown that it was only during the past 10,000 years that Lake Turkana in the northern Kenya was isolated from the Nile system. When connected, it would have been an excellent connection between the heartland of the continent and the Mediterranean.

Another question focuses on the extensive stone-walled villages and towns in Southern Africa. The Great Zimbabwe is but one of thousands of standing monuments in East, Central, and Southern Africa that attest to considerable human endeavor in Africa long before contact with Europe or Arabia. The Neolithic period and Iron Age still offer very great opportunities for exploration and discovery.

As an example of the importance of history, let us look at the modern South Africa where a visitor might still be struck by the not-too-subtle representation of a past that, until a few years ago, only "began" with the arrival of Dutch settlers some 400 years back. There are, of course, many pre-Dutch sites, including extensive fortified towns where kingdoms and nation states had thrived hundreds of years before contact with Europe; but this evidence has been poorly documented and even more poorly portrayed.

Few need to be reminded of the sparseness of Africa's precolonial written history. There are countless cultures and historical narratives that have been recorded only as oral history and legend. As postcolonial Africa further consolidates itself, history must be reviewed and deepened to incorporate the realities of precolonial human settlement as well as foreign contact. Africa's identity and self-respect is closely linked to this.

One of the great tragedies is that African history was of little interest to the early European travelers who were in a hurry and had no brief to document the details of the people they came across during their travels. In the basements of countless European museums, there are stacked shelves of African "curios"—objects taken from the people but seldom documented in terms of the objects' use, customs, and history.

There is surely an opportunity here for contemporary scholars to do something. While much of Africa's precolonial past has been obscured by the slave trade, colonialism, evangelism, and modernization, there remains an opportunity, at least in some parts of the continent, to record what still exists. This has to be one of the most vital frontiers for African exploration and discovery as we approach the end of this millennium. Some of the work will require trips to the field, but great gains could be achieved by a systematic and coordinated effort to record the inventories of European museums and archives. The Royal Geographical Society could well play a leading role in this chapter of African exploration. The compilation of a central data bank on what is known and what exists would, if based on a coordinated initiative to record the customs and social organization of Africa's remaining indigenous peoples, be a huge contribution to the heritage of humankind.

MEDICINES AND FOODS

On the African continent itself, there remain countless other areas for exploration and discovery. Such endeavors will be achieved without the fanfare of great expeditions and high adventure as was the case during the last century and they should, as far as possible, involve exploration and discovery of African frontiers by Africans themselves. These frontiers are not geographic: they are boundaries of knowledge in the sphere of Africa's home-grown cultures and natural world.

Indigenous knowledge is a very poorly documented subject in many parts of the world, and Africa is a prime example of a continent where centuries of accumulated local knowledge is rapidly disappearing in the face of modernization. I believe, for example, that there is much to be learned about the use of wild African plants for both medicinal and nutritional purposes. Such knowledge, kept to a large extent as the experience and memory of elders in various indigenous communities, could potentially have far-reaching benefits for Africa and for humanity as a whole.

The importance of new remedies based on age-old medicines cannot be underestimated. Over the past two decades, international companies have begun to take note and to exploit certain African plants for pharmacological preparations. All too often, Africa has not been the beneficiary of these "discoveries," which are, in most instances, nothing more than the refinement and improvement of traditional African medicine. The opportunities for exploration and discovery in this area are immense and will have assured economic return on investment. One can only hope that such work will be in partnership with the people of Africa and not at the expense of the continent's best interests.

Within the same context, there is much to be learned about the traditional knowledge of the thousands of plants that have been utilized by different African communities for food. The contemporary world has become almost entirely dependent, in terms of staple foods, on the cultivation of only six principal plants: corn, wheat, rice, yams, potatoes, and bananas. This cannot be a secure basis to guarantee the food requirements of more than five billion people.

Many traditional food plants in Africa are drought resistant and might well offer new alternatives for large-scale agricultural development in the years to come. Crucial to this development is finding out what African people used before exotics were introduced. In some rural areas of the continent, it is still possible to learn about much of this by talking to the older generation. It is certainly a great shame that some of the early European travelers in Africa were ill equipped to study and record details of diet and traditional plant use, but I am sure that, although it is late, it is not too late. The compilation of a pan-African database on what is known about the use of the continent's plant resources is a vital matter requiring action.

Vanishing Species

In the same spirit, there is as yet a very incomplete inventory of the continent's other species. The inevitable trend of bringing land into productive management is resulting in the loss of unknown but undoubtedly large numbers of species. This genetic resource may be invaluable to the future of Africa and indeed humankind, and there really is a need for coordinated efforts to record and understand the continent's biodiversity.

In recent years important advances have been made in the study of tropical ecosystems in Central and South America, and I am sure that similar endeavors in Africa would be rewarding. At present, Africa's semi-arid and highland ecosystems are better understood than the more diverse and complex lowland forests, which are themselves under particular threat from loggers and farmers. The challenges of exploring the biodiversity of the upper canopy in the tropical forests, using the same techniques that are now used in Central American forests, are fantastic and might also lead to eco-tourist developments for these areas in the future.

It is indeed an irony that huge amounts of money are being spent by the advanced nations in an effort to discover life beyond our own planet, while at the same time nobody on this planet knows the extent and variety of life here at home. The tropics are especially relevant in this regard and one can only hope that Africa will become the focus of renewed efforts of research on biodiversity and tropical ecology.

An Afrocentric View

Overall, the history of Africa has been presented from an entirely Eurocentric or even Caucasocentric perspective, and until recently this has not been adequately reviewed. The penetration of Africa, especially during the last century, was important in its own way; but today the realities of African history, art, culture, and politics are better known. The time has come to regard African history in terms of what has happened in Africa itself, rather than simply in terms of what non-African individuals did when they first traveled to the continent.

Egyptian Boat on the Nile River, c. 1870 (Hippolyte Arnous).

INTRODUCTION

THE BATTLE OF THE NILE

The sun was descending over the Mediterranean Sea on August 1, 1798. French Admiral François Paul de Brueys peered from his battleship's quarterdeck toward the horizon and got the scare of his life. British warships were entering Abukir Bay, where his thirteen vessels swung at anchor. Brueys had thought his fleet safe and invincible inside this harbor near the mouth of the River Nile.

He was wrong. The masterful English sea devil, Lord Horatio Nelson, was braving the shoals across the bay's entrance to attack the French ships where they lay—despite approaching darkness. The French had not dreamed Nelson possessed such nerve. They were ill prepared to fight what would be perhaps the most decisive naval battle of the eighteenth century.

Earlier that year, Napoleon Bonaparte's French army had arrived in Egypt. In the ongoing power struggle between the two nations, Napoleon's arrival posed a grave threat to the British. The French land forces were roaming the desert sands along the northern gateway to Africa, an unconquered continent with natural riches that lured all the European powers. Meanwhile, Napoleon's navy was prowling the Mediterranean, supporting the invasion of Egypt, and threatening British interests in Africa and the

Middle East. Nelson had chased the French fleet for months. Finally, he had it within his sights. But could he defeat it?

On paper, the battle seemed evenly matched. The two fleets contained the same number of ships, although the French had more cannons. All the ships were manned by seasoned sailors and gunners and commanded by veteran officers.

The French vessels were anchored in a line along the shore. Brueys's captains made ready the guns on the seaward side of their ships. They believed that if the English drew abreast, they could give an excellent account of themselves, broadside for broadside. What they did not count on was British nerve and cleverness. Five of the English warships sailed across the bow of the lead French vessel, then maneuvered into the dangerously shallow waters between the French ships and the beach. Halfway down the French line, their captains ordered the anchors dropped. The rest of Nelson's ships glided halfway down the seaward side of the French line and also stopped. The foremost French ships now were caught in a hellish vice, cannonaded from both sides, while their comrades to the rear were unable to maneuver into the fray and assist them!

After three hours, most of the leading French vessels were sunk or battered beyond usefulness. Hundreds of men were dead or dying, torn apart by grapeshot and flying splinters from shattered decks and masts. Brueys was killed, as were several of his captains. The French admiral's flagship exploded when fire spread to its powder magazines.

Throughout the night, the English pounded the rest of the French fleet. Egyptians flocked to the shoreline to watch the awesome, grotesque fireworks just off their coast. Mangled corpses and broken ship timbers floated ashore. Could it be possible these warring foreigners would destroy each other . . . and leave Egypt in peace?

Dawn, however, revealed that the English, though bruised and weary, had emerged victorious. The British now controlled the Mediterranean Sea, and Napoleon's army soon would be forced to return to France. A strange new era was about to begin for this land of desert and river, ancient marvels and mysteries.

What did this battle between a handful of rickety wooden ships representing the navies of two foreign countries have to do with the way Egypt would be governed a century later? Much! A century later, these same two European powers would be major players in the infamous "scramble for Africa"—a power struggle that would deeply affect every region of the great Dark Continent. By 1798 events were already lining up for the scramble. The countries of Europe were jockeying for control of the most important areas along the African coast.

But what was so important about Egypt, aside from the mythical riches that were rumored to lie hidden inside the great pyramids? The answer was one word, water.

Egypt, as you probably know, consists mostly of desert. It seems hardly the place for explorers to go looking for water. Yet, Egypt attracted the attention of England, France, and other western nations because of three crucial bodies of water found in and around it: the Mediterranean Sea, the River Nile, and the Red Sea.

The Mediterranean, naturally, was the first object of control, for it is the sea that separates the European and African continents. Trading vessels from many lands and cultures have relied on its shipping routes for thousands of years. The Nile—longest river in the world—is the great watercourse entering the Mediterranean from southward, deep in the East African interior. Control of the Nile meant control of territories far beyond Egypt's desert sands.

The third waterway, the Red Sea, was not nearly as important as the other two to European strategists at the beginning of the nineteenth century, but nevertheless held immense promise. This finger of the Indian Ocean between Africa and Arabia was known to lie only 100 miles from the southeastern corner of the Mediterranean. If a canal someday could be dug to connect the two oceans, the shipping route between Europe and the Orient

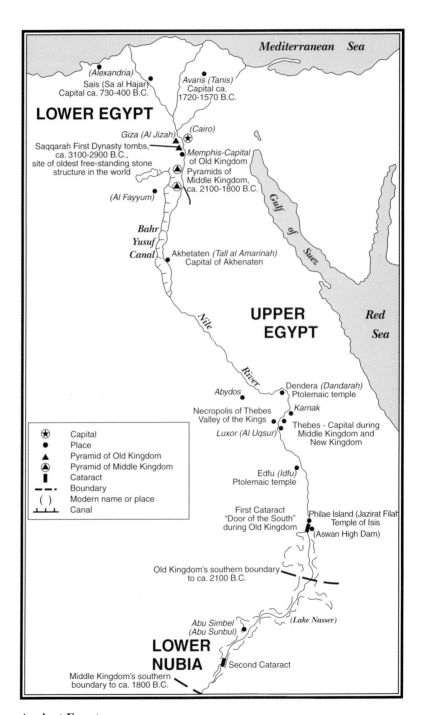

Ancient Egypt

would be greatly shortened. In time, as we will see, the Suez Canal would accomplish exactly that.

These three watery features have shaped Egyptian history, culture, economics, religion, and government for ages . . . but never more than during the past two centuries, the period of growing English domination followed by Egyptian independence.

In the aftermath of Nelson's victory at the Battle of the Nile, England held the key to Egypt. It did not yet dominate this land of the ancient pharaohs. But its influence would be profound—and would increase enormously—during the next century.

A Pottery Shop, Cairo, c. 1870 (G. Lekegian)

FOREIGNERS CONQUER A ONCE GREAT KINGDOM

From the edge of the busy Suez Canal zone today, peering west and south into the barren land, you are struck by the stark difference in scenery. A few short miles take you from an engrossing center of commerce into a barely populated desert. It is odd to realize most of Egypt is, in fact, desert. And the desert does not gradually encroach upon the city; rather, it suddenly and completely overtakes it. As you traverse these few miles from city to wasteland, you are cast from a modern, thriving international trade zone to a bleak, unsettling, almost lifeless void.

Egypt is unique by virtue of its geography and its antiquity. This desert wasteland was the domain of one of ancient history's cradles of civilization and its location has always given it strategic importance in international affairs. Napoleon Bonaparte called Egypt "the most important country in the world" when Europeans converged on the region 200 years ago.

Egypt was a great prize to foreign invaders long before the Europeans arrived. Greek, Roman, and Arab armies roamed its deserts as they began probing North Africa more than 2,000 years ago. They came in quest of new territories, slaves, riches, and useful natural resources that could be transported to their homelands. These invaders intermarried with the natives. Thus, they brought their customs and ideas into Egyptian life . . . and carried elements of Egyptian traditions back to their home countries.

They were overrunning the domain of one of the world's oldest and most fascinating urban cultures. Archaeologists believe humans may have lived in Egypt more than 8,000 years ago. Over the centuries, Egyptian culture influenced other peoples within the African continent—ancient Middle Easterners and Greeks—and since western civilization has been influenced in many ways by early Greece, that means it also has been influenced by Egypt.

The Egyptians, in turn, absorbed some of the customs of the invaders through intermarriage and servitude. Elements of modern Egyptians' ancestry can be traced to Turkey and Rome in the north, Ethiopia in the south. Some modern residents of Egypt, such as the Bisharin, claim Arab ancestry, yet resemble the Egyptians shown in ancient artwork. The Bisharin language, too, is a combination of Egyptian and African dialects.

From ancient times, Egypt has been thought of as a land divided. It seems odd to westerners, but northern Egypt is called "Lower Egypt" and the southern region "Upper Egypt." Bear in mind that the River Nile flows from south to north. Its mouth (the "lower" part of a river) is the Nile delta, spanning many miles of Mediterranean coastline. When historians speak of Lower Egypt, they refer to the delta at the northern coast. Upper Egypt is considered the area between Cairo, the capital, and Lake Nasser at the southern border with Sudan.

Around 3100 B.C., the various peoples of Lower and Upper Egypt were united by a king named Menes into what became a powerful, sophisticated kingdom. Pharaoh Menes founded the city of Memphis as his capital. This marked the beginning of Egypt's long "dynastic period," a *dynasty* being a line of kings descended from the same family.

The Egyptian dynasties were the world's first known form of national government. The ancient Egyptian pharaohs are considered by historians to have been the world's first "kings"—

Rue Nahassin, Old Quarter, Cairo, c. 1870 (G. Lekegian)

rulers of a country, not merely a village. Using, sadly, slave labor, they built awesome structures like the great pyramids—perhaps the most famous of the world's human-constructed wonders to this day. The Egyptian ruling class was wonderfully artistic and highly educated, developing forms of writing and mathematics.

Some thirty Egyptian dynasties ruled the land for almost 3,000 years, from the time of unification under Menes until Alexander the Great of Macedonia conquered the Nile delta in 332 B.C.

Before the invasions of foreign cultures, the ancient Egyptians worshipped many gods. Each city had a god, and pharaohs had their own gods. Menes worshipped Horus, a falcon-like god of the sky, son of the gods Osiris and Isis. Menes believed he and Horus were one and the same, keeping watch over their sprawling kingdom from a lofty vantage point. Egyptians throughout the dynastic period believed their pharaohs were gods.

Ra, the sun god, was worshipped in the city of Heliopolis beginning in the fourth dynasty. Ammon was the god of Thebes; the town and its god came to national prominence during the fifth dynasty. The greatest temple in history was constructed at Karnak near Thebes in honor of Ammon. Eventually, Ammon and Ra became one in the minds of the ancients. After the time of the Romans, Egypt became largely a Christian land. Ancient Christians in Egypt were known as the Copts.

We tend to think of Africa as a continent "to be explored" by outsiders. The notion that Africans themselves have gone exploring might seem odd, but they have indeed! One of the

Old Quarter, Cairo, c. 1870 (G. Lekegian) *Cairo is the largest city in Africa. It has stood for more than 1,000 years at the same site on the banks of the Nile River. Today the extensive medieval section of the city has more than 400 registered historic landmarks including monuments, mosques, buildings, and massive gateways.*

earliest explorers on record, if not the earliest, was an Egyptian named Hannu. He was authorized by King Sahure to lead an expedition from Hamamat, a River Nile port about 400 miles upriver (south) from Cairo.

Hannu's hardy band climbed the Nile cliffs and set out across the eastern desert. Reaching the Red Sea coast, they plodded southeastward along the shore. Near the place where the Red Sea angles into the Indian Ocean, they explored the eastern tip of the African continent, which Hannu called "Puoni," or "Punt." According to his journal, Hannu's objective was to "fetch for Pharaoh sweet-smelling spices." If the accounts are accurate, he also obtained great quantities of gold-silver alloy, a perfume known as myrrh, and a rare type of wood believed to have been ebony.

The most interesting part of the story is that Hannu's expedition took place around 2750 B.C.—40 centuries before European explorers first began seriously probing the African coast. By 1400 B.C., the Egyptian Empire extended the full length of the eastern Mediterranean coast. It encompassed modern-day Israel and reached the lower region of the Hittite Empire (modern Turkey).

Alexander of Macedonia conquered much of the Mediterranean and Asian regions in the fourth century B.C. He made the important port city, which he named Alexandria, the Egyptian capital and appointed one of his generals to govern the country. This man became known as Ptolemy I.

Ptolemy and his successors ruled for 300 years. At certain times, they occupied large parts of Arabia and northern Africa; at other times, they fell under foreign control. The last Ptolemaic ruler, the famous Cleopatra VII, committed suicide in 30 B.C. after an intriguing but futile attempt to divide the powerful Roman invaders and preserve Egyptian rule.

A Fey Shop, Cairo, c. 1870 (G. Lekegian) *In British dialect,* fey *means fated to die; calamity or evil; supernatural; unreal; appearing to be under a spell; and being in unnaturally high spirits. A person having an apprehension that something evil was about to occur would purchase a remedy at a fey shop.*

The Romans occupied Egypt for almost 700 years. For the most part, it was a time of peace. Egypt was an extremely valuable Roman province, rich in minerals for manufacturing and a source of great stores of grain along the Nile Valley. Alexandria became a center of culture and a hub of commerce, bringing together merchants from the northern Mediterranean ports, the African continent, the Arab countries and even India. Naturally, it became a melting pot of different peoples.

With brief interruptions, the Romans controlled Egypt until A.D. 639. In that year, Arabian invaders defeated the weakened remnants of the dying Roman Empire and brought Egypt under Muslim dominion. Though the changes since then have been many, Egypt remains a Muslim nation to this day.

It was a fierce army of 4,000 Arabians who crossed the Isthmus of Suez into Egypt. Sent by the Muslim caliph of Umar, they set about occupying the land. At the important battle of Heliopolis in 640, they defeated the Byzantine (Eastern Roman Empire) army. Two years later, the Arabs had spread throughout the Lower Nile region.

This was the beginning of a complex sequence of Muslim regimes that would shape the character of modern Egypt. Most of the native Egyptians who worshiped their ancestral gods eventually converted to the Muslim religion. The Coptic Christians living in Egypt at the time of the Muslim conquest were allowed to practice their faith for awhile, but after a revolt in A.D. 829–30, they were defeated and persecuted. As a minority faith, Coptic Christians survived and can be found in Egypt

Bisharin, 1914 *The Bisharin are a nomadic people who live a poverty-stricken existence in the desert valleys of southern Egypt. Bisharin tend herds of camel, goats, and sheep. They also trade senna leaves, once used in medicines, which they collect in the desert. Though claiming Arab descent, the Bisharin bear a strong resemblance to the surviving depictions of predynastic Egyptians. Their language is a curious combination of Egyptian and East African dialects.*

Temple of Horus, Edfu, c. 1870 (J. Pascal Sebah) *The Temple of Horus was unearthed by the French archaeologist Auguste Mariette in the 1860s in Edfu, on the west bank of the Nile near Aswan. (Mariette also suggested the plot for Giuseppe Verdi's opera* Aïda.) *The Temple was then seen in a wonderful, almost perfect state of preservation in spite of the 2,000 years that had passed. Unfortunately, the faces of the kings and gods in the reliefs were scratched out during the Christian period.*

still today. For the most part, though, Egypt had become a Muslim land.

A colorful variety of warring Arab powers—the Turks, the Fatimids, the Syrians, and others—succeeded one another in controlling Egypt for the next thousand years. Perhaps the most interesting were the Mamelukes, freed Turkish slaves who rose to power in 1250 after helping repel the European crusades. They made Egypt a renewed center of Mediterranean power over a period of three centuries. The Mamelukes were great warriors. It was they who stopped the Mongol horde's advance into southwestern Asia in 1260. But they also were important patrons of literature and architecture, and they oversaw a flourishing era of Egyptian trade.

The Mameluke era came to an end in 1517 when Selim I, sultan of the Turkish Ottoman empire, conquered the Mamelukes and all their holdings in Syria and Egypt. From then until the coming of the Europeans, Egypt was ruled by a succession of Ottoman viceroys, or *pashas,* during a period noted for its lack of cultural or social advances. But Mameluke descendants continued to hold positions of power in government, and by the time of Napoleon's arrival in the late 1790s, the Mameluke elite had regained their sovereignty over the Lower Nile valley. Mameluke and Turkish leaders struggled for dominion over the country.

This tension was the excuse Napoleon used to invade Egypt. He allied himself with Turkey and pledged to wrest control of Egypt from the Mamelukes and restore the country to the Ottoman fold. His army defeated the Mamelukes at the Battle of the Pyramids near Cairo. The French emperor, of course, had designs of his own on the Nile. He briefly established a French dictatorship at Cairo. But after Nelson's naval victory in 1798 and Napoleon's later failure to conquer Syria, the French withdrew. They left Egypt at the beginning of the 1800s in the hands of the British and the Turks.

It was to be a very stormy century in Egypt, and the first lightning strikes flashed immediately. The Turkish army sent to subdue the Mamelukes in Egypt consisted in part of Albanian soldiers. They rebelled and, led by an Albanian-born soldier named Mehmet Ali, later called Muhammad Ali, joined the Mamelukes in ousting the Turks.

Muhammad Ali in 1805 became *pasha* of Egypt. He was a brilliant leader, but ruthless. He literally seized control of the country. When the leaders of his Mameluke allies protested his sovereignty, he had them slaughtered. The climax of his bloody power play came in 1811. He summoned two dozen Mameluke leaders to a court ceremony . . . where he ordered their assassination!

Muhammad Ali was intent on building a formidable army in order to retain power. He also wanted to modernize the country, based on the European model. During his long reign, he placed the daily running of society largely in the hands of Egyptian natives. Natives were conscripted, or drafted, into the army to fight their leader's many wars. (Some natives crippled themselves to avoid military service.)

Meanwhile, Muhammad Ali looked to Europe for ideas. He brought European military, agricultural, and educational advisers into the country and sent many Egyptians to Europe for training.

Ironically, Muhammad Ali recognized the authority of the Ottoman sultan. He was willing to be known as the *khedive*— the Ottoman's designated governor—of Egypt. During the next decade, he helped the Ottoman Turks strengthen their grip over the Middle East by sending forces against Turkey's enemies. At the same time, he extended his own power by sending an army southward into Sudan. There, he took control of the slave trade. He used some of the slaves to expand his military machine.

While exerting force to expand and secure his domain, Muhammad was intent on making Egypt a healthy country. He began organizing and improving its agriculture by importing new products such as cashmere goats and silkworms. The introduction of long-staple cotton production during his reign gave Egypt a farming resource that has served the country well to the present day.

Edfu, c. 1870 (Hippolyte Arnous)
Travellers were advised to avoid lingering in Edfu except for renting a donkey in order to visit the Temple of Horus. The one "hotel," upper left, was described as "primitive and yet somewhat expensive."

Philae, c. 1870 (J. Pascal Sebah) *The ruins on the island of Philae were partially submerged by the old Aswan Dam in 1900. The temples reemerged severely damaged in the 1970s with the completion of the Aswan High Dam. The island was then leveled and the temples rebuilt. The formal reopening took place in 1980.*

The original island of Philae was about 500 yards in length and 160 yards in breadth. With its stately temples and rich vegetation, it had been one of the most beautiful points in Egypt. The temple of Isis was the principal sanctuary—part of which is pictured on the left of the photograph. The building on the right is the later-built Greco-Roman kiosk of Trajan.

Although Muhammad Ali was a strong leader for almost half a century, his improvements for Egypt ultimately came to little. Some historians conclude he attempted too many tasks in a land that could not support them for the long term. To his great dis-

Temple of Osiris, Abydos, c.1870 (J. Pascal Sebah) *Abydos, a most ancient and sacred city, is located in the low desert west of the Nile. It was a royal necropolis and then a pilgrimage center for the worship of Osiris, the most important god of the ancient Egyptians. Around and between the various temples at Abydos is a vast complex of cemeteries used in every period of Egyptian history from the prehistoric through the Roman era.*

appointment, many of his new factories and schools eventually had to be closed.

After Muhammad Ali died in 1849, his descendants governed an Egypt that officially was under Turkish control but was tied increasingly to European governments. The British were allowed

Muhammad Ali Mosque, Cairo, c. 1870 (Hippolyte Arnous)

Polishing Silver, c. 1870 (G. Lekegian) *These men are polishing silver in front of the Muhammad Ali Mosque. Its lofty and graceful minarets are one of the most conspicuous sights in Cairo. Completed in 1857, the mosque is part of a walled fortress called the Citadel.*

to establish a national telegraph company and a national bank. The French were granted the right to build the Suez Canal. Our next chapter will examine how this engineering marvel, completed in 1869, would affect Egypt's entry into the twentieth century and its journey toward independence.

Ismail Pasha, Muhammad Ali's grandnephew, became viceroy in 1863. The line of power normally would have passed to Ismail's older brother Ahmad, but Ahmad was killed aboard a train that plunged into the Nile in a mysterious—and suspect—railway accident. Some of Ismail's relatives believed he had something to do with the tragedy, but it was never proved.

Tilling the Soil Using an Ox and a Dromedary, c. 1870 (G. Lekegian)

Pyramids and Sphinx, c. 1870 (Hippolyte Arnous)

Ismail had been educated partly in Austria and France. He, even more than Muhammad Ali, longed to modernize Egypt and Sudan, based on the inventions and advances he'd seen in Europe. More railroad tracks were laid. Canals were dug. New schools were built. But Ismail also focused vainly on attractive but basically useless and costly "improvements." Public streets were given facelifts, palaces and monuments were built. These cosmetic projects were intended to impress visiting foreign dignitaries.

Ismail's innovations and luxuries cost enormous sums of money. Soon Egypt's government was deep in debt to England and France. In 1875, he was obliged to sell his government's interest in the Suez Canal to England, and the stage was set for the Europeans to descend upon Egypt in force.

Before we conclude our review of pre-European Egypt, it's worth remarking that the Egypt of ancient times left us one of the most fascinating studies in human history. Much physical evidence exists to tell us about life during the early dynasties and afterward, as Roman and Arabian powers took control of the Nile region.

Unquestionably, the most easily recognized symbols of ancient Egypt are the great pyramids. Almost 5,000 years old, they are the oldest known human-made stone structures in the world. More than 80 pyramids and similar-shaped buildings are found along the banks of the River Nile. Giovanni Belzoni, an Italian archaeologist, in 1818 explored the pyramid of Chephren at Giza. To his despair, he found it had been looted, possibly centuries earlier.

To archaeologists and linguists, one of the most important finds of all time was the Rosetta Stone. During Napoleon's failed invasion of Egypt in 1799, French soldiers found the slab of black basalt near the Egyptian town of Rosetta, not far from the Mediterranean coast. It proved invaluable to scientists because it contains text in three different languages: Greek, demotic, and hieroglyphic. Since scholars already were familiar with Greek, this stone thus held the key to deciphering other ancient passages—especially those written in hieroglyphs. Egyptian records from about 3000 B.C. until the coming of the Romans typically were recorded in hieroglyphs, a system using pictorial letter characters.

By 1822, largely through the studies of French scholar Jean François Champollion, hieroglyphic writings were plainly translatable. These translations shed light on other ancient writings. Egypt, to Europeans, became more than just another intriguing foreign place; it became a key to understanding primitive human history.

The Nile's importance to historians and archaeologists, however, meant little to the western governments who arrived during the 1800s. To them, Egypt was a key not to understanding our past, but to controlling the continents of Africa and Asia.

Arab Sheik, Suez, 1899 *Suez is located at the southern end of the Suez Canal. An ancient Red Sea trading site, the city was revitalized with the opening of the Suez Canal (1869). The 1908 Baedeker's* Egypt *notes that prior to the Suez Canal, Suez "was a miserable Arabian village, while in 1897 it contained 17,457 inhabitants, including 2,774 Europeans. Its trade, however, in spite of the opening of the canal, and the construction of large-scale docks, has not materially increased. Neither the Arabian quarter, with its seven mosques and unimportant bazaar, nor the European quarter, which contains several buildings and warehouses of considerable size, presents any attention. The streets and squares are kept clean, and the climate is excellent."*

This photograph was taken by Captain M.S. Welby (1866–1899). Welby, one of England's most promising younger explorers, was killed in the Transvaal during the South Africa War. His extensive trips through Africa and Asia, especially to northwest Tibet, were vividly reported in the popular British press. The Royal Geographical Society's obituary notice commended his "fairness and tact in dealing with native races, which enabled him to pass unscathed, where many would have met with obstruction and violence."

2

THE CANAL

F rench engineer Ferdinand de Lesseps and his crews found themselves laboring in a most interesting place, geologically. To the east, the dry, rugged highlands of the Sinai peninsula opened into the continent of Asia. To the west was the Nile delta, gateway to Africa.

They were not the first workers to build a canal in this region. In the nineteenth century B.C., Egyptians had dug a channel for irrigation that eventually linked the Nile basin with the isthmus lakes and even with the Red Sea. The Roman conquerors improved it and named it Trajan's Canal after one of their emporers. It functioned for about 2,500 years, until Arab armies covered it up in A.D. 775.

When European mariners began rounding the Cape of Good Hope at the bottom of Africa to trade in the orient, they lamented the long, perilous route. Surely, they believed, it would be worth the effort to connect the Mediterranean and the Gulf of Suez with a canal across the narrow isthmus. This could save them months off each voyage. In fact, the Suez Canal eventually would shorten that trip by 5,000 miles!

Cairo, c. 1870 *Tourism to Egypt expanded enormously after the opening of the Suez Canal in 1869. Many came on their way to India. Thomas Cook & Son's opened a Cairo office. Carl Baedeker published* Egypt *and by 1908 Baedecker's guide was in its sixth "remodeled" edition—with more than 400 pages of tourist information and suggested itineraries. "A glimpse of the country may be obtained in 4 or 5 weeks," recommended Baedecker, "2 or 3 days may be devoted to* Alexandria *and the journey thence to* Cairo, *10–12 days may be spent in* Cairo *and its neighborhood in the manner suggested, 3 days may be given to the* Fayum *[funerary portraits dating from the Roman period], and 14 days or more may be devoted to* Upper Egypt *(railway to Luxor or Aswan) while a few days must be set aside for resting." Burton Holmes, the travel writer and lecturer, regaled his American audiences with stories of Egypt's glamour and excitement.*

At this point, photography was in its infancy—but several photographers realized the financial potential of combining their skill with this surge in tourism. Their photographs, both the local scenes and the posed studio ones, were mounted on a stiff cardboard and sold at numerous tourist kiosks. Their work is the only surviving record we have of them—G. Lekegian; Hippolyte Arnous; Felix Bonfils; Emile Bechard; Antonio Beato; J. Pascal Sebah; and Zangaki. Were these their real names? What were their nationalities? Ironically, the Royal Geographical Society has an extensive collection of their realistic and detailed photographs. But the society's records are mute about who donated them or when they were donated.

Not until Napoleon's occupation of Egypt was a survey of the isthmus made, however. Napoleon's engineer incorrectly concluded the sea level was different by 33 feet in the Mediterranean and the Gulf of Suez, meaning locks would have to be built. Such an undertaking would have been far more formidable in 1800 than several generations later. No matter; Napoleon was obliged to withdraw from Egypt before he could plan a canal project.

The French did not give up the idea, however. In 1854 de Lesseps obtained permission from Said Pasha, the khedive of Egypt, to build a canal. Several years later the khedive granted France an equally important concession: The Compagnie Universelle du Canal Maritime de Suez (Suez Canal Company) was organized, owned by French and Egyptian interests. This company was organized with $40 million in capital—an astronomical sum in the nineteenth century. It was entitled to operate the planned waterway for ninety-nine years. After that, control of the canal was to be turned over to Egypt.

Interestingly, de Lesseps did not select the shortest path through the Isthmus of Suez. As an engineer, though, he found the sensible one. A few miles of approach channels had to be dredged in the floor of the Mediterranean Sea at Port Said in the north and in the Gulf of Suez (a finger of the Red Sea) on the southern end. Altogether, the canal is 101 miles long.

Many canals, as mentioned earlier, include a series of locks. These are controlled sections of the watercourse in which the surface level can be raised and lowered. They are needed if the sea level at one end of the passage is not quite the same as at the other. Locks adjust the ship to the correct sea level as it passes through the canal. As de Lesseps discovered, the Suez Canal would require no locks because the surface level of the Mediterranean and Red seas is approximately the same. The canal would be an "open passage." It would run basically straight, but with a number of notable bends. At several places, it would open into broad lakes. The largest of these is *Al-Buhayrah al-Murrah al-Kubra,* the Great Bitter Lake, about 20 miles above the southern end of the canal.

Work began in 1859. Using 25,000 native peasants—paid, but toiling under conditions hardly improved from ancient slave labor—the company excavated with picks and shovels, carrying away the sand and soil in baskets. Later steam-powered machines did the bulk of the digging and dredging. Bad weather, deadly diseases, and unrest among the laborers slowed progress. The project had been expected to take six years to complete; it took ten. It cost about $100 million—an unimaginable sum, to people of that day.

In August 1869 the work was finished. Ismail Pasha wanted to impress the Europeans with an extensive grand opening celebration. The famous Italian composer Giuseppe Verdi was commissioned to write an opera based on an ancient Egyptian story, to be performed at a fine new opera house at Cairo. (Thus was written the classic *Aïda*—but because the costumes weren't completed in time, *Rigoletto* had to be substituted as the opening night performance.) Visiting royalty from the great nations raised their eyebrows when they saw the new palaces and brightly lit streets Ismail had ordered installed in the major cities.

The first vessels passed through the canal in November 1869. From that time on, the old shipping route between Egypt and the Orient, around the Cape of Good Hope, rarely would be used again for commerce.

At first, the canal was open only during daylight hours. Ships with proper lighting were allowed to move through the water at night beginning in 1887. Naturally, the canal was built to accommodate the ships of the day. Soon, vessels were being built larger, and the canal had to be widened. When it was completed, the canal was 230 feet wide at the surface, tapering to a depth of 72 feet at the bottom. It was 26 feet deep. These measurements reflected the comparatively modest dimensions and cargo capacities of mid-nineteenth-century cargo ships. Today the waterway is much larger: more than 700 feet wide at the surface and about 300 feet wide at the bottom. It is 64 feet deep. Thus, it can handle much bigger ships with greater drafts.

The canal has undergone other modern improvements to meet ever-growing shipping needs. In 1986 Egypt opened the Damietta port complex on the northern end of the canal. Damietta is set up to handle 16 million tons of cargo.

About 60 vessels pass through the Suez Canal daily. It takes approximately 15 hours to navigate from end to end. Ships are directed in a single lane for most of the passage, but two lanes operate on the large lakes, and elsewhere they can use passing bays. A railroad operates along the western bank of the canal. At the time it was built, the village of Suez was the only notable settlement along the canal. Many towns later grew on its banks, for the Suez Canal quickly became one of the most heavily trafficked shipping lanes on the globe.

When Ismail Pasha found his administration shackled by debt, he reluctantly sold the Egyptian shares in the Suez Canal to England in 1875. The canal was managed by a commission consisting primarily of British and French appointees. The Egyptians were not happy with the arrangement, but Ismail had no alternative.

Thus, almost from the beginning, there has been tension over control of the all-important canal. At times the result has been dangerous military confrontations—despite an international convention that in 1888 guaranteed all nations the right to use the Suez Canal. The countries of the world immediately recognized the unparalleled importance of this canal, situated in the middle of the world, connecting the continents.

Soon after the canal was completed, the major powers of Europe began the "scramble" for Africa that we discussed earlier. This complicated jockeying for territory would be tragic (but at moments almost comical). Amid all the positioning and power plays on the Dark Continent, it is especially interesting to note that Egypt came under the thumb not of France, but of England, even though France was the country that obtained Egypt's permission to build the Suez Canal.

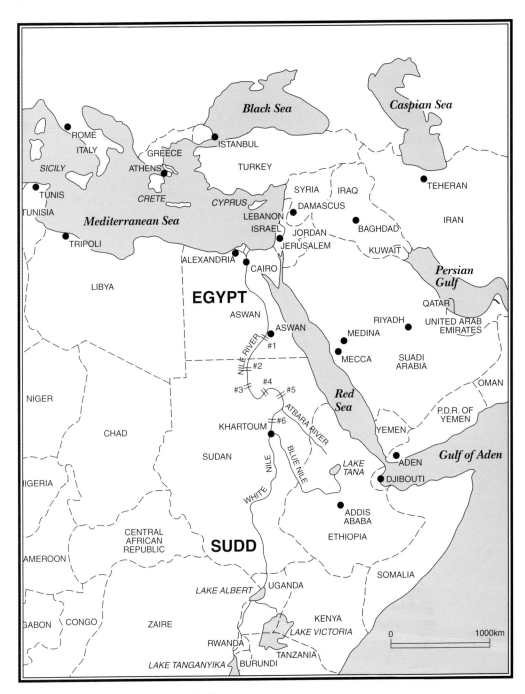

North Central Africa and the Mediterranean

Indeed, when de Lesseps had sought funding for the project from English investors, they had flatly refused. The British thought the canal would be so expensive to build it never would pay for itself. Lord Henry Palmerston, a noted Victorian statesman, vowed to "oppose the work to the very end." Even after the canal opened, the British postal service for two years maintained the Victorian attitude of aloofness, claiming canal travel was too slow for official use. A few years later, of course, the British were singing a very different tune.

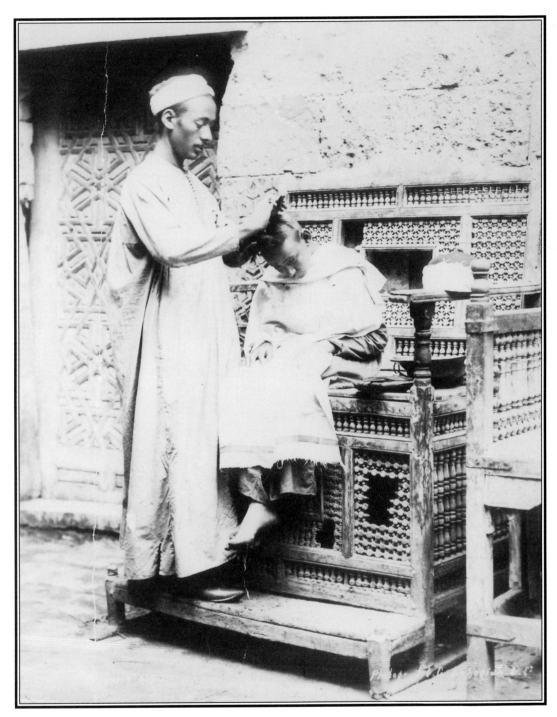

Barber Shop, c. 1870 (G. Lekegian)

3

WEST MEETS EAST IN THE SHADOW OF THE PYRAMIDS

Money equals power in international relations. Ismail Pasha, Egypt's khedive from 1863 to 1879, needed money to pay for the vast improvements he had in mind for Egypt. European investors had money to lend, and doing so would obviously give them influence in this country that was so strategically important, being situated at the juncture of the African and Asian continents. In time, these loans of money gave Europeans actual control over certain aspects of Egyptian government.

Foreign creditors weren't Ismail's only money problems. Several years after he became the Egyptian khedive, the country's farming economy suffered a severe crisis. Its roots lay in an unlikely place far away: the American South.

"Cotton was king" in the South, as the saying went. But with the outbreak of the American Civil War, cotton production in that region fell off sharply. Egypt was well positioned to help supply the void in world cotton demand. Years before, under Muhammad Ali, long-staple cotton had become a flourishing crop along the Nile. Suddenly, Egyptian traders ecstatically found they could command four times the prewar price for their cotton.

Basketmakers, c. 1870 (G. Lekegian) *These basketmakers of Cairo seem to be using sewed-braid coiling. In this method, sewing is done with a needle or an awl which binds each coil to the preceding one. This type of basket existed in ancient Egypt—and is still used throughout North Africa.*

What goes up, of course, must come down. The Civil War ended. Soon America was producing cotton once more for its own textile mills and those of Europe. Egypt was not prepared for the drastically tumbling prices that resulted. Its economic plight worsened.

In time, cotton would become more important than ever among Egypt's economic assets. Long-fiber cotton from Egypt has supplied English clothing factories for more than a century,

Grocer's Shop, Cairo, c. 1870 (G. Lekegian) *The cylindrical items at the top of the photograph are chunks of sugar in a sealed paper wrap. Many governments placed a tax on sugar sales—and this may have been true in Egypt of the 1870s.*

Woodturner, c. 1870 (G. Lekegian)

Coppersmith, Cairo, c. 1870 (G. Lekegian) *This photograph was taken in the busy Sharia en-Nahhasen, with its many coppersmith souks. To the right is a wall of the Dar Beshtak palace, which was constructed c. 1330.*

and Egypt produces more of it today than any other country. For a time, though, the cotton crash of the 1860s dealt a harsh blow to Egypt's economy.

Ismail's careless spending habits were accompanied by an excess of ambition. He wanted to be not just another Ottoman viceroy controlled by the sultan in Istanbul, but the sovereign of an independent country that was growing in importance. To the Europeans, whom he wanted so much to impress, he tried to present Egypt with its new canal as a major member of the world community, not simply a piece of the Ottoman Empire.

Fellahin Woman and Child, c. 1870 (G. Lekegian)

Irrigation Apparatus, c. 1870s (G. Lekegian) *The* shaduf, *pictured here, is a primitive apparatus used in irrigation. It consists of a bucket and a counter-weight. It is set in motion by one person only and the water is drawn into buckets that resemble baskets in appearance. As in this picture, several* shadufs *are sometimes arranged one above the other. In order to distribute the water equally, the flat fields were divided into a number of small squares using a few inches of earth as a barrier. These could be easily opened or closed so as to regulate the water within them.*

Irrigation Apparatus, c. 1870 (Hippolyte Arnous) *Another* shaduf, *or hand-operated device for lifting water, invented in ancient times is shown here. The operator pulls down on a rope attached to the long end to fill the bucket and then allows the counterweight to raise the bucket.*

Naturally, the Ottomans frowned on this attitude. Although, for the most part, Ismail was on cordial terms with the sultan in Istanbul, there were strained moments. And in the end, when Ismail's financial bind with his European creditors brought him to the end of his rope, the sultan would not stand behind him.

The Egyptian khedives, beginning with Muhammad Ali, had encouraged European involvement in the country's development. Foreigners were engaged to build factories, train soldiers,

Water Boring Tools, Dakhla Oasis, 1927 *The Dakhla Oasis is in the western Egyptian desert near Libya. The area is an almost rainless plateau. The fertility of an oasis is based on the availability of fresh water from small springs. Shown here are tools used to bore for water.*

and teach college students. Under Ismail Pasha, the pace quickened. Mile after mile of new railroads, telegraph lines, and irrigation canals were built. Street lights and water mains were installed in the cities. By 1875 Ismail had placed his country in debt by about a million English pounds. His solution, as we saw in the last chapter, was to sell Egypt's stock in the Suez Canal to Great Britain.

Egypt lost more than control of the canal. The British, French, Italians, and Austrians created a special financial office to collect Egypt's national debt over a period of years. Most of the country's tax revenues began going toward debt reduction—and the Europeans began taking significant government positions in Egypt. It was becoming obvious to Egyptian citizens that foreigners, not the khedive, were running their country. And the foreign administrators were not always honest in their dealings. They engaged in smuggling and money-lending at exorbitant interest rates while dodging local taxes.

Egyptians of all classes disliked the Europeans. The people who suffered worse at the hands of the foreigners, as you might expect, were the peasant workers, the *fellahin*. They were forced—sometimes brutally—to work harder than ever to produce crops and other goods for paying the interest demanded by European creditors.

Ismail scrambled to find ways to pay down the national debt and to unshackle his government from European control. His position was hopeless. Facing unbearable pressure from both the European creditors and his own people, Ismail was deposed by the Turkish sultan and went into exile in 1879. His son, Tawfiq, succeeded him. Tawfiq plainly was a pawn manipulated by the Europeans. To take back their country, a number of native military and business leaders formed the National Popular Party and gained control of the government. Angry citizens supported them with riots protesting foreign dominion.

Ahmed Arabi (Arabi Pasha), an army colonel, led this early nationalist movement. In September 1881 he and other army officers led a military coup against the khedive. Their primary objective was simply to win improvements in the lot of the native officers. Arab officers had been protesting wretched pay and second-class treatment, compared with British officers. The Ottoman leadership had ignored them. Now they demanded to be heard.

Arabi soon found himself representing more than just disgruntled soldiers. A broad segment of the whole Egyptian populace was unhappy with the foreigners. Perhaps, they thought,

Arabi's "demonstration" against the khedive could lead to independence! For months, the Egyptian administration was held in limbo. Militant nationalists worked the country into a virtual powder keg of angry discontent.

It exploded in June 1882. Citizens in Alexandria took to the streets in a bloody riot, attacking European residents and their friends and burning property. Among the 50 people killed was the British consul.

England responded with force the following month. A naval bombardment destroyed much of Alexandria. Thousands of British marines came ashore there and along the Suez Canal. Armed with rapid-firing Gatling guns, the British defeated the Egyptian army in only two months of fighting. In this manner, Egypt was added to the British Empire.

While the brief war between Britain and the Egyptian rebels brought bloodshed and long-lasting animosity, one humorous story emerged. Spying on a British encampment, the Egyptians noticed people dressed in skirts. They assumed this was a tent city inhabited by wives and girlfriends of the redcoat soldiers. If they could attack and kidnap the women, the Egyptians believed, they would hold a valuable bargaining chip.

In reality, the skirted campers were the kilt-wearing Gordon Highlanders from Scotland—some of the fiercest soldiers ever to fight for Queen Victoria. A night attack by the Egyptians led to disaster. The Scots sent them reeling. It is said that the Egyptian commander was so demoralized at being defeated by "women" that he ended the revolt.

What would Great Britain do with Egypt now that it held military control? Many statesmen in London, including Prime Minister William Gladstone, did not really want England to be "in charge" of Egypt. They feared it would rankle other European powers, particularly France, as well as the Ottomans who still claimed the land.

Evelyn Baring, who later would be named Lord Cromer, was the statesman sent to Cairo to take charge in 1883. He would remain in that post almost twenty-five years. As the British

Egyptian Woman, Cairo, c. 1870 (G. Lekegian)

consul-general to Egypt, his first objectives were to stabilize Egypt's economy and calm the masses.

Baring was appalled at the economic condition of the country. He realized Ismail's borrowing policies had ruined Egypt. At the same time, he scowled at the European creditors and merchants, including his British countrymen, who had taken advantage of Ismail's predicament. In a biting commentary, he stated, "Egypt must have been an earthly paradise for all who had money to lend at usurious rates of interest, or third-rate goods of which they wished to dispose at first-rate prices."

Eventually, the Egyptians should be able to resume their old system of government under the Ottoman khedive, and the British could go home. Or so many British leaders assumed. In fact, however, the English wouldn't be leaving for three-quarters of a century.

The khedive never officially lost power. He remained the Egyptian figurehead, although he dared do little without British approval. Most historians regard the khedive Tawfiq, Ismail Pasha's son, as a barely disguised puppet of the English consul-general. They generally agree, though, that without British military might behind it, the Ottomans' khedival form of administration could not have survived—which Tawfiq well understood. Tawfiq undoubtedly resented the presence of Baring and the British armed forces in his country, but he needed them.

Baring was an intellectual and an aristocrat, but also a tireless worker and demanding leader. He seems to have given little thought to the plight of Egypt's longsuffering fellahin class. He made almost no effort to establish a bond with the Arab population; he didn't bother to learn Arabic, the language of the land. More pressing, to him, were the problems of securing Britain's hold on the country—especially the canal. Baring appointed more and more British bureaucrats to government posts formerly held by native Egyptians.

For appearances, the British set up an Egyptian parliament. Native ministers were appointed to certain government posts, and Tawfiq remained in place as the khedive. But only the British had real power.

Tawfiq was succeeded in 1892 by a young nationalist-minded khedive, Abbas Hilmi II. Abbas wanted to untie the strings by which Great Britain manipulated Egypt. He was no match for Baring, though. As khedive, he was little more independent than Tawfiq had been.

And Egypt still needed British military muscle. As we'll see in the next chapter, Muslim rebels had driven the Europeans from Sudan in the mid-1880s. The insurgents had grown more powerful and threatened to invade Egypt. Many Egyptians feared their country easily would be overrun without a powerful British army present. Even with British soldiers on hand, there was a sense of uneasiness.

Although the British administrators did little to advance health care and education among the natives, they brought about certain improvements. Baring by the 1890s had accomplished his primary goal of getting Egypt out of debt. Sadly, he did it not so much with brilliant business policies as with heavy taxes on the fellahin. To his credit, however, he increased Egypt's agricultural output by improving irrigation systems.

Their newfound freedom from the national debt led many Egyptian natives to decide it was time for the British to withdraw its government bureaucracy, if not its military support. Baring was inclined to withdraw neither. As he grew older, he seemed more intent on establishing the British permanently in Egypt, the gateway between England and its most important colony, India.

Remember that the European powers by now were in the process of "carving up" the African continent. England, France, Italy, Germany, and other countries were laying claim—both informally and with a series of peculiar, one-sided treaties—to different regions, especially around the coast and along the major rivers.

In North Africa, France was trying to control Algeria, Tunisia, and Morocco. It eventually let Libya go to the Italians and—in an amazing bit of diplomatic blundering—let Egypt go to England without serious contention. When Egypt and Sudan were wracked by internal revolts during the 1880s, the French were too busy elsewhere to interfere. England took the opportunity to put down the revolts and take control over Egypt. It had the right to intervene, the British government reasoned, since the unrest in Egypt and neighboring Sudan prevented Egypt from paying off its national debt to the Europeans.

After 1883 Baring's system of rule became known as the "Veiled Protectorate." To the world, it appeared Egypt was governed by Egyptians. The British, however, were the decision makers—or, as they preferred to be remembered, the "protectors"—of a country whose affairs had gone badly awry. Many local leaders were eager to cooperate with the British occupiers, realizing resistance would be futile. The working class, though, faced heavier burdens than ever before. From the outset of the British era in Egypt, resentment was building.

Bedouin Doctor, c. 1870 (G. Lekegian) *This Bedouin doctor was photographed wearing his regalia and full tribal outfit.*

4

"Sharing" the Government: The Condominium Era

The history of Egypt is linked closely to the history of Sudan, its neighbor to the south. Muhammad Ali had brought much of Sudan, the largest country in Africa, under his influence during the first half of the nineteenth century. To the British, controlling Sudan was almost as important as controlling Egypt. Via the Nile, Sudan is the geographical border between the Egyptian deserts and the interior of Africa. But Britain hardly knew how to deal with Sudan, which presented its own set of problems to any European nation that might seek to rule it.

Ismail Pasha in the 1860s and 1870s sent explorers into Sudan and beyond, looking for the source of the Nile. He also dispatched soldiers to put down Sudan's highly active slave trade. Ismail hoped these missions would serve two purposes: (1) satisfy the Europeans' humanitarian sentiments and (2) extend Ismail's control over much of Sudan.

This was a risky venture, though. Ismail knew it would anger powerful Sudanese as well as some of his own countrymen. Egyptian traders were among the profiteers of slaving expeditions in the interior. Ismail, however, bowed to the influence of the British, who opposed slavery.

In Sudan, as in Egypt, natives resented the overbearing presence of the British. The British-led forces approved by Ismail to thwart slave raiding infuriated important tribal leaders, who derived much of their wealth and power from the slave trade.

Under the Muslim religious leader Muhammad Ahmad, known as "The Mahdi," in the early 1880s, Sudan revolted. Mahdist forces in November 1883 slaughtered a British-led Egyptian army at the Battle of El Obeid. Worse was to come. A popular but reckless British general, Charles George Gordon, was sent the following year to lead the remaining Egyptian-British forces in a general withdrawal from Sudan. The British government had decided Sudan for the moment was too hot—and too large—to hold. Gordon apparently thought otherwise. Rather than withdraw, he and his soldiers occupied Khartoum (today the capital of Sudan) and quickly found themselves under siege by the Mahdists.

The record is unclear whether Gordon deliberately took Khartoum in hopes of breaking the Mahdist rebellion or was trapped there beyond his control. Whatever the case, he gradually realized his position was grave, and he begged London to send a force to relieve him. The British government and military were slow to respond . . . too slow. When reinforcements finally trudged southward from Egypt after many months of stalling, Gordon's food and supplies had dwindled to nothing. His small force could put up little resistance. The Mahdi's hordes overran Khartoum in January 1885 and massacred Gordon and his entire command.

In England the public was outraged. For the time being, though, England was not prepared to subdue Sudan and avenge Gordon. When vengeance finally came, it was decisive. An army led by Major General Horatio Herbert Kitchener in 1898 methodically put down the various rebel forces in Sudan. England now had another huge territory to add to its empire . . . and to spread its foreign armies and administrators still thinner.

The British knew they would never really "control" Sudan. They hoped simply to keep the peace while preventing other European powers, most notably France, from infiltrating that part of Africa. To maintain even a low-key government in Sudan, they needed help from the Egyptians. They also needed

Fellahin, c. 1870 (G. Lekegian) *Perhaps nowhere is the relationship between the people and the Nile River so intense as in the floodplain areas. Here, the population can survive only by making the most careful use of the available land and water. These Egyptian fellahin (peasants) lived near Aswan. Their descendants had to be relocated when the High Dam was completed in 1970.*

to create the appearance that they had not "conquered" the Nile regions but merely were "protecting" them. An element of native control had to be made visible to the world. In 1899 the British government contrived a peculiar arrangement for overseeing Sudan this way. Called a "condominium" government, it united English and Ottoman authorities. The Egyptian khedive, who traditionally answered to the Ottoman Turks, would continue to be the official ruler of both Egypt and Sudan. British

Arab Sheik, c. 1870 (G. Lekegian) *A sheik is the chief or the head man of a village or tribe.*

agents and Egyptians would share the actual administration of the government.

Some of the Egyptian administrators were given meaningful positions at fairly high levels of government. The critical roles, however, were reserved for the British. It was obvious to observers of the "scramble for Africa" that England effectively had won Egypt and Sudan.

A legion of British bureaucrats now steamed for the Mediterranean. What they found along the Nile was a land wholly unlike their native England. "Nothing in this strange land is commonplace," commented Lord Alfred Milner, an Anglo-Egyptian administrator in the early 20th century.

Egypt may have appeared strange to the wide-eyed English officials and their families, but it was little different from the land that other ambitious foreigners had encountered for centuries past. The awesome pyramids and dynastic artifacts, the life-giving Nile, the desert vistas unoccupied by any human or animal to every horizon—all this had been witnessed by Macedonian and Napoleonic soldiers. The peasant folk, too, the fellahin, lived, dressed, and labored almost exactly the way their ancestors had before them. A journalist visiting the country in 1922 wrote:

> When you see the modern fellah at work with mallet and chisel, or scratching the sun-baked plain with his crude hoe, or dipping his clumsy fish-net into the Nile, he is, in face and physique, startlingly like the pictured Egyptians of the Pharaohs' times.
>
> Since prehistoric days this race, a vast farming colony, has lived along the Nile and in that great delta which ages of floods have built out into the Mediterranean. Though the Persian conquest . . . ended the period of native rule, the mental and physical aspects of the modern fellah are, so far as we can judge, exactly like those of his early ancestor who sweated under the Pharaohs. . . .

Culturally, the fellah has been Arabized; he speaks a form of Arabic and turns to Mecca in his prayers. Otherwise he is the same silent, melancholy, inscrutable person who doggedly dragged granite blocks for hundreds of miles to build the pyramids, who blindly bent to the big sweeps of the early Egyptian galleys, or who conceived and began to dig the Suez Canal centuries before de Lesseps was born.

Three primary racial groups comprise the Egyptian population in modern times. Most prominent are the Hamito-Semitic people. Their ancestors were the dwellers of the Nile valley for thousands of years.

An interesting minority within the Hamito-Semitic race is the Berber people. They live in the great desert far west of the Nile, near the Libyan border. Siwa Oasis is the center of Berber life, although the Berbers are spread through other villages, as well. They speak a unique language and cling to customs that seem odd, both to foreigners and to other Egyptians. Berber women go to special lengths to shield their appearances; their clothing covers practically their entire bodies, with only a slit across the face revealing the eyes.

A second major racial group is the Arabic people. When Arabs invaded Egypt in the seventh century A.D., they literally came to stay. Arabs today are one of the nation's three largest races. They include about half a million Bedouin nomads. Life has changed little for many Bedouins during the centuries of

Bedouin Chief, c. 1870 (G. Lekegian) *The Bedouins are an Arabic-speaking nomadic peoples of the Middle Eastern deserts. While they make up but a small part of the total population, they utilize a large part of the land area. Following World War I (1914–1918), the Bedouin tribes had to submit to the control of the countries in which their wandering areas lay. Today the Bedouins of Egypt live in the western desert and are either seminomadic or totally sedentary.*

Bedouin from Sinai Peninsula, c. 1870 (G. Lekegian)   *The Sinai Peninsula is an area between the Gulf of Suez and the Suez Canal on the west and the Gulf of Aqaba and the Negev Desert on the east.*

foreign domination over Egypt. Although some Bedouins have adopted urban lifestyles, many still live in animal-hide shelters near desert oases. The Victorian English administrators saw them—as we still see them today—wearing several layers of long garments, their faces partly concealed by colorful cloths, which shield them from the withering sun. They herd and sell livestock for a living.

Other wandering peoples are scattered through the Sinai region, who are not Bedouins but are descended more recently from the neighboring Arab lands.

The Nubian people, Egypt's third major race and its largest non-Arabic group, have roots in southern Egypt and northern Sudan. This is now the region of Lake Nasser. Tens of thousands of Nubians were forced to abandon their Nile villages and relocate when the great Aswan High Dam was finished and the lake waters rose. Like the Berbers and Bedouins, the Nubians—a tall, thin, dark-skinned race—continued to live like their forebears until recent times. They transported goods up and down the river. They also farmed and fished for a living. During the twentieth century, the majority of Nubians moved to Egypt's cities.

The British found dwelling in Egypt the descendants of other invading groups from history. These included Turks, Ethiopians, Greeks, Romans, and other Europeans. The Greeks, in particular, comprised a large foreign contingent who lived in Egypt well into the twentieth century.

Life in Egypt and Sudan during the condominium era was fairly peaceful compared with the region's long legacy of conflict. Not all was well, however. Native Egyptians' resentment never diminished toward the British intruders, who had assumed most of the power and much of the country's wealth for themselves. That resentment gradually grew into a widespread spirit of nationalism. Throughout the period of British involvement, urgings for independence created periods of tension, sometimes violence. Advocates for independence staged demonstrations against British policies. They organized strikes by the work force.

Among those emboldened to call for home rule was a Muslim religious leader named Muhammad 'Abduh. As a young scholar in Cairo during the 1870s, 'Abduh began following the teachings of Jamal ad-Din al-Afghani, a charismatic Islamic teacher from Persia who had established himself in the Egyptian capital. Because of his revolutionary politics, the Egyptian administration deposed Afghani in 1879.

Despite his allegiance to the exiled Afghani, 'Abduh became editor of a government-run newspaper. He promoted reforms in Muslim practices and in the country's social conditions. He also openly criticized the Europeans who had infiltrated and dominated the Egyptian government. In the 1882 uprising, which brought the British army into the country in force, 'Abduh was suspected of being one of the revolutionary conspirators. The British expelled him from the country.

'Abduh went to Paris, where Afghani was publishing an underground political journal. After working briefly with Afghani, he moved to Beirut, Lebanon, to teach at an Islamic college. Then his life took a very different course. In 1888 the Egyptian government not only allowed him to return to his homeland but appointed him to a judgeship.

He rose in his judicial career until he ultimately was appointed *mufti* of Egypt. The mufti was the country's chief Islamic legal authority. 'Abduh obtained this post largely through British influence in 1899. By this time, he was urging native Egyptians to cooperate with the British to achieve social reforms. This shift in attitude apparently stemmed from his conclusion that improvements in education and living conditions were more important than national self-control. 'Abduh was one example of the turn-of-the-century Egyptian educated class who determined that opposition to the powerful English would be useless. Although they obviously did not like foreign domination over their country, they knew the time was not right for an independence movement to succeed.

The mufti worked for the education of children living in poverty. However, he attracted more attention—and ire—by pressing for changes in Islamic law. For example, Muslims

Nubian Girl, Luxor, 1914 *Nubia was a region of ancient Africa. It extended along the southern boundary of ancient Egypt to present-day Khartoum in the Sudan. The Egyptians obtained gold from Nubia. The Nubian kingdom lasted until about A.D. 350. During the 500s the Nubians were converted to Christianity. In the 1300s, when Arabs conquered this area, the Nubians became Muslims.*

were forbidden to eat meat that had been processed by non-Muslims; 'Abduh considered such rules backward. So from his appointment as mufti until his death in 1905, 'Abduh was at odds with both Muslim traditionalists and advocates for Egyptian independence.

Bad feelings between elitist British occupiers and native workers turned tragic in incidents such as the 1906 Dinshwai conflict. Peasants living at Dinshwai, a Nile delta village, earned part of their living by selling pigeons. Understandably, they were angered by English army officers who came once a year to shoot the birds for sport. At last, some of the peasants attacked the officers on their pigeon-shooting holiday and beat them with sticks. One Englishman ran back to the nearby garrison for help. He died, either from heatstroke or from a head wound—probably both.

Punishment was swift and deadly. Four Dinshwai villagers were executed and many more flogged and imprisoned. When word of the episode reached London, some members of the House of Commons expressed shock at the harshness of the punishment. But in the minds of British military and government leaders in Egypt, such measures were necessary to keep the natives in line. Dinshwai and other incidents, however, had a backlash effect, making more and more native Egyptians determined to see an end to foreign domination.

The British-Ottoman arrangement for controlling Egypt and Sudan ended with the outbreak of World War I in 1914. The Ottomans, remember, were Turkish. Turkey sided with Germany—the enemy of England and France. When Turkish and German military forces threatened the Suez Canal, Britain announced lone control over Egypt. In December 1914 the British leadership declared Egypt a "protectorate" and sent Abbas, the khedive, into exile. They made his uncle, Hussein Kamil, the new puppet ruler and gave him the grand-sounding title of *sultan*. Quite unlike the famous sultans of Turkey, though, Hussein Kamil exercised little control over his country.

For the first time in almost 1,300 years, Egypt officially was free of Arab control. The change hardly mattered to most of the

Olive Press, c. 1910 *This primitive screw type press dates back to the ancient Egyptians and probably has not changed that much through the centuries. About two-thirds of the olive production of Egypt is pressed into olive oil.*

Arab Sheik and His Servants, 1904 *This photograph was taken in southern Egypt by George Hughes. Hughes, a cartographer, prepared detailed maps showing the mining concessions granted to various companies by the Egyptian government. His maps, which were published in 1904, are still useful to archaeologists because they include the location of ancient mines, wells, and caravan routes.*

people, though. They had borne the English yoke for more than 30 years. The fellahin now were set to manual labor for the war effort. The British army commandeered their livestock for food. Ousting the khedive would not improve the natives' quality of life, and it certainly did not mean self-government was at hand.

Arab Women and Children, Cairo, c. 1870 (G. Lekegian)

By now, of course, the population had become thoroughly Arabized. The great majority of Egyptians felt a closer kinship to the Ottoman Empire than they did to Great Britain. The Arab religion and way of life had become so ingrained that when Egypt finally did cast off all outside authority and become a free republic 40 years later, it would become a free *Arab* republic.

England and France, aided by the United States, defeated Germany and its allies in 1918. Turkey had lost its gamble in siding with the Germans and would never return to power in Egypt. An Egyptian figurehead still held forth in a royal palace, but English authorities made the major decisions. It was a land of growing restlessness, though. Egypt would not remain long within the British Empire.

Street Scene, Aswan, c. 1920

5

A CONTINENT
CLAMORS FOR
SELF~RULE

During the first half of the twentieth century—especially in the years between the two world wars—a mood for independence was growing all across the African continent. By 1950 it was obvious to most European power brokers that they couldn't keep their African colonies, protectorates, or puppet kingdoms much longer. They would have to recognize self-ruling governments. The best they could hope for were peaceful transitions and friendly relations with the new independent nations. That way, they could arrange valuable trade agreements and continue to benefit from such African exports as oil, mineral ores, and farm crops.

In Egypt the natives' passion for independence was fueled by factors unique to that country. In its beginnings, the movement toward an independent Egypt was not a grass-roots effort that built its strength among the common people. Instead, it was an idea entertained mainly by middle-class native merchants and military officers. By the end of the First World War in 1918, however, it appealed to a broad segment of Egyptian society, from the lowly fellahin to the well-to-do.

Nationalist leaders in Egypt organized the Wafd party immediately after World War I. (The word *wafd* is Arabic for "delegation.") They agreed not to interfere with British control over the Suez Canal, but otherwise, they demanded an independent Egypt. When the British refused, the nationalists began staging demonstrations against the protectorate administration.

Great Britain's response: Jail the Wafd leaders.

The native Egyptians' rejoinder: Revolt!

In the spring of 1919 mass demonstrations, strikes, and clashes with British police and soldiers threw the country into a crisis. Nationalists cut telegraph lines and severed train rails. Egyptian officials within the government went on strike. Business effectively shut down. Inevitably, hundreds of people, mostly Egyptians, were killed.

The turmoil lasted almost three years before Great Britain reached an agreement with the Wafd leadership. In 1922 Ahmad Fuad, the country's sultan (basically the same title as that of the former khedive), was named King Fuad I. Egypt thus became a kingdom and took its place in the world as a separate country, in concept at least. It had its own constitution. A governing legislature with a senate and a lower house of deputies was established. Egyptian nationals were allowed to vote for its members.

National Geographic magazine in October 1922 reported on "Egypt's daring and successful drive for independence." The British, it said, had "relinquished their protectorate." The United States, for one, now recognized "the new nation on the Nile." At the time, that summary seemed accurate. The period of the "protectorate" indeed was over. As for real independence, though, time would tell a different story. Several critical areas of control were left to the British. They included defense, the administration of Sudan, the protection of foreign residents and property within the country, and the protection of minorities.

The recognized head of the Wafd during these events was Saad Zaghlul. In a sense, Zaghlul was the first major leader in the drive for an independent Egypt. His parents were Nile peasants who fared well for their class. He was able to attend Al-Azhar University in Cairo and even to go to law school. By 1892 the young lawyer had become an appeals court judge. Soon afterward, his social standing improved still further when he married a daughter of the country's Ottoman prime minister.

In the early 1900s Zaghlul became involved in government and politics. He served as the Egyptian minister of education from 1906 until 1910, then served two years as minister of justice. He helped found the People's Party, composed of native Egyptians who generally favored cooperation with British officials.

Shortly before the beginning of World War I, Zaghlul was elected to Egypt's Legislative Assembly. There he began voicing criticism of British policies. As a result, he became a prominent figure in the growing nationalist movement. The assembly was disbanded when Britain severed ties with the Ottomans and deposed the khedive in 1914. Although the country saw little political activity during the war years, Zaghlul and others increasingly advocated independence. They feared Britain's tight wartime control over the Suez Canal and the Nile delta would lead to Egypt becoming a formal colony if Britain and its allies prevailed against Germany and Turkey.

Britain was, as we know, victorious. Within days of the peace signing, Zaghlul led a Wafd delegation in a meeting with Britain's high commissioner in Egypt, Sir Reginald Wingate. They issued their courteous but firm call for England to end its protectorate regime. Wingate predictably refused, with the predicted results. With his arrest, Zaghlul became a hero to the Egyptian people. When the nationalists later obtained their concessions from England in the early 1920s, the Wafd became the dominant native political party, and Zaghlul was named prime minister.

It was a time of progress toward independence, but not toward peace. Fanatical factions assassinated British officials. Most notably, nationalists in Cairo murdered Sir Lee Stack, governor-general of Sudan, in November 1924. Under the threat of a severe military response from London, Zaghlul, by then approaching seventy, resigned.

Although the Wafd in 1922–1924 won a measure of self-government for Egypt—on paper, at least—British soldiers still occupied the canal zone and surrounding country. The canal was, as we have seen, one of the main reasons England ever involved itself in governing Egypt. An 1888 international convention had given every country on earth the right to send ships through the canal, whether in peacetime or war. Britain, however, held firm military control over the canal.

During World War I, British troops prevented enemy ships from passing through the waterway. Two decades later, with Hitler's Germany rising to power in the 1930s, a nervous Great Britain again felt a need to secure the canal.

A journalist named Maynard Owen Williams, traveling through the canal in 1935, wrote, "The Suez Canal is a lockless sand ditch connecting two landlocked seas and three lakes. . . . But with industrial Europe at one end and the populations and raw materials of the East beyond, this sand ditch is a barometer of world life."

With the creation of the Egyptian "kingdom" in the 1920s, three strong, separate forces began to influence Egypt's policies. The British administration remained the ultimate source of power. The Wafd party was its native adversary, fanning the flames of independence all along the Nile and in the port cities. Between them, aligning himself sometimes with one side, sometimes with the other, was King Fuad.

During the years between the world wars, the Wafd was the dominant political party in Egypt. In fact, in the unsettled aftermath of World War I, it was the natives' only party of significance. Zaghlul gave it dynamic leadership and drew to the Wafd a growing mass of supporters. In the country's first legislative elections, held in 1924, the Wafd became the majority party in office. By the time Zaghlul died in 1927, however, the Wafd was being fractured. Some of its members, impatient for independence, formed more aggressive minority groups. They disapproved of Wafd leaders negotiating with the British on matters of national policy. To them, the withdrawal of the British from Egypt was the only issue of real importance.

It seems amazing to many Americans that foreign countries like Egypt can make much headway when countless political parties are jockeying for power and pulling against one another.

They forge temporary alliances—occasionally with their staunchest rivals—in order to outmaneuver certain opponents on certain issues. They sometimes stage election boycotts. Not infrequently, they subdivide into even more factions. Violence among them is not uncommon.

In contrast, in our country it seems sometimes enough of a task to distinguish the positions of our two long-dominant parties, Democrats and Republicans. When an independent or third-party candidate mounts a significant challenge in some isolated campaign, it is big news!

By the late 1930s Egypt's political tapestry was becoming quite complex. Power struggles within power struggles already had resulted in small offshoot parties asserting political identities. Groups who opposed the established Wafd party included a Fascist youth party, the Greenshirts of Young Egypt, which would support Nazi Germany in the approaching Second World War. The Wafd was losing its dominance, although in various forms it would continue to be an important party throughout the twentieth century.

In 1936 an Anglo-Egyptian treaty gave Britain the right to station armed forces in the Suez Canal zone and the port of Alexandria. This treaty incensed Egyptian nationalists, who intensified demands that the British withdraw from the zone.

The following year brought several events of special significance for Egypt. For one thing, the country became a member of the League of Nations. This membership enhanced its image in the eyes of the world as a self-supporting country.

In May England gave up one of the powers it had retained when it allowed Egypt to become a kingdom in 1922. A new treaty acknowledged that Egypt, not Britain, was responsible for protecting international citizens in Egypt, as well as their property. Such a treaty may not have raised many eyebrows around the world, since in most countries, we naturally assume international visitors are in the hands of that country's native government, in terms of obeying the laws and relying on the police for protection. In the odd relationship between England and Egypt, however, the change was significant. It diminished

Great Britain's potential excuses for intervening with military force in Egypt.

Also in 1937, King Farouk, Fuad's son, took the throne. Farouk was young and eager to exert control. He soon ousted the country's prime minister from office. Although he wasn't a particularly successful problem solver or leader, Farouk's feistiness in the face of the British military presence was an ominous signal to the west. Step by step, Egypt was moving toward real independence.

Egypt was a crucial British ally in World War II. The country was important not just because of the Suez Canal, but because of its established British military bases. Germany invaded North Africa—a military prize because of its oil fields. In the early years of the war, British and German forces fought desperately across the desert dunes. Not surprisingly, most Egyptian nationals did not energetically support England's war effort. To them, this was another struggle among the long-hated European powers. The German army's presence on the continent obviously was unwelcome, but hardly less welcome, in the minds of some Egyptians, than the British. In light of Germany's initial desert victories, many natives expected Hitler to win the war.

Some of Egypt's military leaders actually favored the German-Italian-Japanese Axis. So did certain officials in King Farouk's court. They believed the Egyptian people had more in common with the Axis nations than with the western Allies. The British in Egypt had to keep an eye on their native support base while battling the German Afrika Korps.

Once more, however, England and the Allies prevailed against the German-led powers. Immediately after the war, the Allies spearheaded the formation of the United Nations—the world body that was hoped to be a more effective keeper of international peace than its predecessor, the League of Nations. Significantly, Egypt was a founding member of the UN.

Egyptian Prime Minister Mahmud Nuqrashi shortly after the end of World War II called on Great Britain to pull out its forces from all of Egypt. As in the past, London refused. And as before, the nationalist movement—by now strong throughout the land—encouraged demonstrations and strikes. Many in England knew the time was near for their forces to vacate the land of the pyramids. Withdrawal was not a cut-and-dried issue, though. Control of the Suez Canal no longer was Great Britain's driving motive for remaining in Egypt. (In fact, Britain's need to monitor and police the canal passage diminished markedly in 1948 when India, its great eastern colony, became an independent republic.) Now some British leaders were feeling a *moral* obligation to remain posted in Egypt. Without the stabilizing effect of the British military, they realized, Egypt and other countries in the region might be at violent odds.

The Jews, who at the time were struggling to form the State of Israel, were one example. Another was Egypt's neighbor to the south, Sudan. Sudanese nationalists had their own independence movement on the brink of fruition. Some British officials in the region believed Sudan in fact was ready for independence. Many Egyptians, however, opposed independence for Sudan as a separate nation. They wanted to see Sudan join Egypt as a single Nile republic. A minority of Sudanese nationalists agreed with them; most didn't. A forced merger undoubtedly would have led to bloodshed.

Zeal for an independent Egypt intensified. The country was boiling with political initiatives from both the left and right wings. Communist groups were increasingly active, and by now, Muslim factions with their broad regional religious issues had become part of the Egyptian political fabric. These forces were more militant than the Wafd, which had coexisted in relative peace with the British administration. A group called the Muslim Brotherhood held bloody demonstrations in Cairo and ultimately engaged in outright terrorism. Political leaders on all sides lived in danger of assassination.

The Muslim Brotherhood was one of the most notable of the growing political parties. Organized in 1928, it demanded not simply British withdrawal and independence. It wanted Egypt to be governed by Islamic law—a true, self-proclaimed Muslim nation.

Already, Egypt had become a key member of the Arab League, organized at the end of World War II. The pressing—and extremely dangerous—issue of the period among the Arab nations was the so-called "Palestinian question." Palestine—the ancient Holy Land also known as Canaan—has been a region long wracked by tension and violence between its Jewish and Arab populations. It is the region on the eastern coast of the Mediterranean Sea that today essentially encompasses Israel and part of Jordan. A United Nations committee in 1947 recommended that the land be divided, part of it going to Arabs and part to Jews. The Arab League rejected the plan. Its countries opposed the formation of a Jewish nation in their midst. They also resented Great Britain for playing what they believed was a key role in pressing the U.N. for Jewish statehood. In 1948, Arab soldiers from Egypt, Syria, Lebanon, and other countries invaded the lands claimed by Jewish settlers.

It was the first of many conflicts that plague life in the region to this day—and threaten global peace, because of a complex weaving of alliances involving the world's superpowers.

The war ended quickly. After two aborted cease-fires, Israeli forces took the offensive, invading Egyptian territory. They overwhelmed their numerous but ill-prepared antagonists and forced a truce. Thus was born the State of Israel, although many Arabs even now refuse to recognize its right to exist.

Unrest continued in Egypt. British soldiers along the Suez Canal found themselves in a state of guerrilla warfare. The violence reached fever pitch early in 1952, when nationalists went on a rampage in Cairo. They burned British property and fought with soldiers. Hundreds were killed or wounded before the city returned to a nervous state of calm.

Britain's worldwide empire was foundering. India, as we've mentioned, led by Mahatma Ghandi and Jawaharlal Nehru, had gained independence in 1948. Now Egypt, land of the all-important Suez Canal, clearly could not be controlled much longer.

The man who would wrest Egyptian independence from the British once and for all came to power six months after the 1952

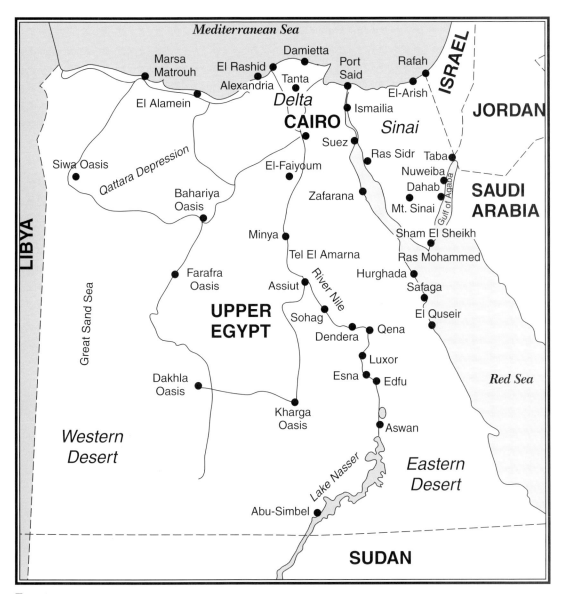

Egypt

Cairo riots. Gamal Abdel Nasser was an Egyptian army commander who secretly helped organize a military group called the Free Officers. In a quiet coup in July 1952 they ousted King Farouk. They threw out the constitution that had been established

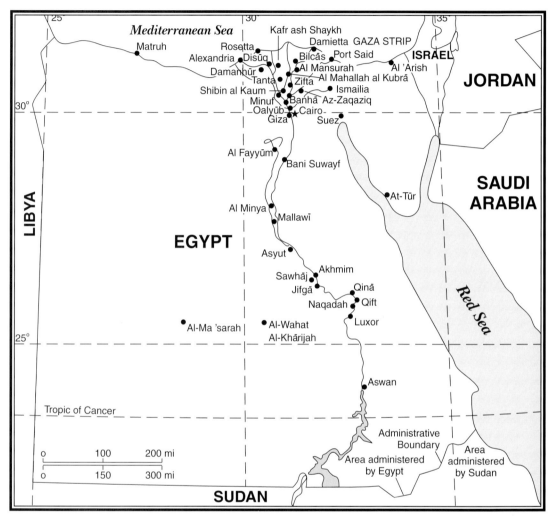

Current day Egypt

30 years before when Egypt became a kingdom, and they temporarily banned all political parties. Egypt was no longer a kingdom but, for all practical purposes, a republic.

Nasser did not want to establish official independence in a costly war against England. Rather, he chose diplomacy. The handwriting was on the wall: Egyptians insisted on governing themselves. Surely, Nasser knew, London could see it.

Finally, the British government acknowledged it could hold the canal no longer. In 1954 a new agreement provided for the gradual withdrawal of English forces. Within two years, the British had turned over their military posts along the canal to Egypt. Egypt formalized its new government in 1956. Nasser was named prime minister and soon afterward was elected president, without opposition.

Nasser was not unchallenged in his rise to power, however. Another member of the Free Officers, General Mohammad Naguib, posed a serious obstacle at first. Officially, Naguib served as the new republic's first president, although his moment of glory was fleeting. He won the support of many members of the diverse political parties that had been outlawed, including the militant Muslim Brotherhood. In September 1954 a Muslim Brotherhood member tried but failed to assassinate Nasser.

Historians believe Nasser privately was not nearly as anxious to destroy Israel, the Arab nemesis, as he was to develop his own needy country. The task of stabilizing a free Egypt, Nasser knew, could hardly be achieved in a state of warfare. Nasser struggled to craft a balanced image: aggressive Arab leader in the eyes of the Arab League; rational, cooperative world leader in the eyes of the world. With hostilities imminent between Egypt and Israel, it would be a very difficult undertaking.

In this setting, the political doors were open to true independence . . . after 2,000 years of rule by foreigners.

The Market, Aswan, c. 1920 *Aswan is on the east bank of the Nile just below the First Cataract. In the town, both Egyptian and Sudanese articles are for sale. The market is renowned for its fine pottery, which is the principal export item, and ostrich feathers. This photograph was taken by Elizabeth Ness (1880–1962), the first woman council member of the Royal Geographical Society. In 1953 she endowed the Mrs. Patrick Ness Award to be presented by the Royal Geographical Society "either to travelers who have success-fully carried out their plan, or to encourage travelers who wish to pursue or follow-up investigations which have been partially completed." She described her own extensive travels in* Ten Thousand Miles in Two Continents *(1929).*

6

A STORMY
INDEPENDENCE

Egyptian leaders during its half century of independence have faced serious ongoing challenges. The problems stem partly from economic weakness, partly from political and religious strife, and partly from the country's geographic location.

Beginning with Nasser's administration and continuing today, the government has had to deal with tension over the ongoing "Palestinian Question" and the presence of the State of Israel. Many Egyptians want to live peacefully. Radical Muslims in Egypt and other Arab countries, on the other hand, demand that Israel be destroyed. Egypt's leaders must struggle to keep the peace in the region while adhering to Islamic ideals and objectives.

At the same time, Egypt is hampered economically. Apart from oil and a few lesser products, it can produce goods of import value only along its waterways. Although Egypt is a relatively large African country, only about five percent of its land mass is habitable and productive. In this sense, its geographic setting works against it.

Happily, Egypt's geography also gives it three unique assets: those bodies of water we studied at the beginning of the book. The Mediterranean makes Alexandria a vital connecting port for commerce into and out of Africa. The River Nile is a highway for this commerce; it also is the great support vein for almost the entire agricultural and industrial output of the country. And the Suez Canal literally brings the whole world to Egypt's door.

The Egyptian government has used the Suez Canal to national advantage on more than one occasion. Shortly after the nation of Israel was formed and became Egypt's arch-enemy in 1948–1949, Egypt forbade the passage of ships to and from Israeli ports. Then, in 1956, the canal became the focus of an international crisis. In July of that year—the month after Great Britain withdrew the last of its forces from the canal zone—President Nasser ordered Egyptian forces to seize the canal. He declared that Egypt would charge tolls for passages and would use the income to pay for construction of the Aswan High Dam on the River Nile.

The announcement shocked the world. It was not a complete surprise to those who had been watching events in the region, though. International politics had come to bear on the Middle East. Religious tension between Israel and the Arab League was complex and dangerous enough. Now there were wider complications—the world superpowers had a stake in Middle Eastern affairs.

Nasser had expected to receive money from the United States and Great Britain to help pay for building the Aswan High Dam, but those two countries decided not to lend their support. They were displeased by Nasser's close relations with the Soviet Union and its communist satellite countries such as Czechoslovakia. The Soviet bloc was giving Egypt weapons of war and providing the country with military advisers. Western powers feared Nasser's Egypt was becoming more like a dictatorship than a democracy. In response to this denial of funds, Nasser announced the canal takeover and tariff.

The Sinai desert and peninsula lay between Israel and the Egyptian mainland. Barely habitable, except on its coasts, it seemed a strange place for a military crisis that would put the whole world on edge. It was a No Man's Land, a desolate buffer between two nations at odds. Egyptian and Israeli forces posted in the region had been harassing each other's positions in undercover raids and open skirmishes. Full-scale warfare seemed imminent.

At the end of October, Israeli soldiers suddenly swarmed across the Sinai and the Isthmus of Suez and invaded Egypt. The Israelis justified their assault on the grounds that Arab guerrilla units hiding out in the Sinai had been attacking Israeli troops and possessions—which was true.

French and English forces joined the fray. Their intent— at least, their official position—was not to support one side over the other, but to guarantee free passage for all ships through the canal. Nasser, in response to the Europeans' interference, ordered several dozen ships sunk in the canal, making it impassable.

Nasser's action seemed reckless to many worried observers in the United States and elsewhere. Conceivably, the deteriorating Suez conflict was moving the entire community of nations toward World War III. Egypt had developed strong ties with the Soviet Union; Israel enjoyed friendly relations with the western countries. If the world's nuclear powers were prompted to take active sides over Suez, who could tell what would happen?

Nasser, however, was reacting largely in his own country's interests. He was a leading supporter of the concept of a regional "Arab nation," a confederacy of Arab-controlled countries in the Middle East and Africa. The British and French, who still had important interests in the region, considered such a union dangerous. To them, Nasser was becoming a radical head of state. He, in turn, suspected the westerners of instigating the Israeli invasion. Some of Britain's own government officials criticized their country for what they saw as a heavy-handed strategy in the Suez crisis.

The Egyptian president's accusations seemed to be proved correct when the British and French air forces based in the Mediterranean flew missions over Egypt in early November. They bombed Egypt's airfields and destroyed its Russian-built warplanes. Egypt was seriously shackled and unable to conduct air raids against Israel. Tension mounted: Would the Soviet military machine come to Egypt's aid?

Happily, a United Nations truce soon was established. The invading armies were withdrawn, international demolition specialists removed the obstructing shipwrecks, and the canal was reopened in 1957. The United Nations posted a peacekeeping force in the area to keep the canal open and thwart Egyptian-Israeli hostilities.

During the Six-Day War between Israel and Egypt in 1967, the canal again was blocked by sunken ships. This time, it remained closed for eight years. That war began after mounting tensions with Israel led to an alliance between Egypt, Syria, and Jordan. When Egyptian troops threatened to cut off Israel's access to the Red Sea, Israel launched a surprise attack that humiliated all three opponents' armies and air forces. Israeli troops once more swept across the Sinai to the Suez Canal.

Nasser, devastated by the defeat, offered to resign. He was encouraged by popular opinion to remain in office, though, until his death in 1970. Anwar el-Sadat succeeded him. Sadat, a former army colonel, had served as Nasser's vice president.

A few months after the Suez Canal was reopened in 1975, Egypt for the first time allowed the passage of nonmilitary ships to and from Israel. In 1979, an Egyptian-Israeli treaty permitted unrestricted shipping through the canal by Israeli vessels.

Egypt's other great waterway, the River Nile, has posed challenges of its own to the young republic. It always has been the lifeblood of Egypt and Sudan. It was so important to the ancients that they regarded it as a god. With twentieth-century technology, Egypt's leaders believed they could derive much more from the river.

The reason Egyptians and their foreign overlords ever wanted to dam the ancient river was to regulate the water supply year-round. For thousands of years, the Nile has changed drastically with the seasons. In early summer, heavy rains in the hills of Ethiopia swell the upper Nile. Until modern times, this created a regular flood on the lower Nile, which ancient Egyptians called the Inundation. Water levels on the river through

Egypt rose as high as 16 feet. The period of high water on the Nile continued well into October.

Then the water level dropped. By March or April—the planting season—the Nile was at its low mark. In years of drought, this delicate cycle became a deadly crisis. Crops failed. People starved.

A dam on the middle Nile, engineers calculated, would provide a sizeable lake. This lake would fill during flood season. The waters could be partly impounded over the winter months, then released for early summer irrigation during the next year's dry season. Later, government officials saw another reason for damming the Nile. With the rapid development of electricity in the early twentieth century, rivers became vital sources for generating hydroelectric power.

The British designed the first dam at Aswan. It took more than four years to build and was completed in 1902. At the time, it was an engineering marvel, although by today's standards it was not outstanding: 90 feet high, just more than a mile across.

The original Aswan Dam helped control the flow of the Nile's water and silt through most of the length of Egypt. By the mid-1900s, though, the Egyptians realized they could get much more benefit from the Nile. Thus was conceived the Aswan High Dam, about four miles upriver from the earlier structure.

This undertaking became a symbol of Russia's world power. Designed by Soviet engineers, it is more than two miles long and 364 feet high. It was completed in 1971 after more than eleven years' labor. The result was Lake Nasser, extending upstream more than 300 miles into northern Sudan.

Located as it is in the desert, Lake Nasser is not a particularly efficient reservoir. Much of its water evaporates before it can be used. Still, the massive project was a vital development for the young republic. Soon Egypt was realizing incredible benefits. Almost a million acres of desert in the vicinity of the Nile now could be watered and made tenable by farmers. In fact, certain crops could be grown all through the year. Fresh water from the Nile also was piped into the arid Sinai peninsula, passing beneath the Suez Canal. Meanwhile, water flowing through the dam became a major supplier of hydro-electric energy. The dam today generates about a fourth of Egypt's electricity.

The cost was enormous, however—not just in money (more than $1 billion), but also in the project's threat to human and wildlife habitats and to national archaeological treasures. As the dam was being planned, it was obvious that temples and other important monuments from the dynastic period stood on sands that would be covered by Lake Nasser. As thousands of workers toiled to build the dam, others toiled—not totally unlike the slaves who erected the monuments many centuries before—to dismantle the gigantic stones, move them to new locations, and put them back together. It was a conservation project unlike any the world ever had seen. Many people from foreign countries joined in the race against time to save the ancient temples.

Also displaced, as we saw earlier, were many people of the Nubian race, whose ancestry long had lived in the borderland between Egypt and Sudan. Furthermore, controlling the Nile led to complications with the Nile's ecosystem. From the time of the earliest civilizations in the Nile valley, the rich silt flowing downstream from the interior highlands has fertilized the whole length of the valley. Now the silt collects in Lake Nasser. Not nearly as much is carried downstream into the lower Nile region.

Gamal Abdel Nasser was known as a social reformer, but his reforms were costly. For example, his educational improvements resulted in a burgeoning number of students . . . with few jobs awaiting them when they completed their studies. The solution—creating new bureaucratic jobs—was counterproductive to the country's economic health.

Egypt's problems proved too large and complex for Nasser to solve. He was able to increase industrial output substantially, but not farming produce. He redistributed the land, reducing the holdings of large landowners and improving the lot of the fellahin. But any gains that were made seemed to be offset by the demands of a growing population.

Ka Oracle, c. 1870 (Hippolyte Arnous) *The* Ka *was prominent in the religious beliefs of the ancient Egyptians. A kind of guardian spirit, the* Ka *accompanied an individual through life as a "double," and after that person's death, continued to protect him against enemies in the next world.*

After Egypt's stand against the western powers in the 1956 Suez crisis, the republic was highly regarded by other Arab nations. It was so influential, in fact, that two countries, Syria and Yemen, briefly united with it to form the United Arab Republic. Nasser was made president of the combined territory. The union was dismantled in 1961, however, because of Syrian natives' frustration with the Egypt-centered government.

Nasser also made overtures to other Arab countries to form an "Arab nation." Several neighboring republics did, in fact, sign a pact in the 1960s, but it had little significance. The Arab countries have much in common, but they also have individual problems and objectives. They have not been willing to accept any head of state as a pan-Arabian leader. Nasser perhaps was the most likely candidate ever for such a role. He was distrusted in certain Arab countries, though: Jordan and Saudi Arabia in the Middle East and, to his west, the North African states of Tunisia and Morocco.

Similar divisions have prevented strong, lasting alliances between Egypt and the African continent's other new nations. Any hope for a pan-African union seems doomed because of age-old differences, notably religion. The Muslim countries in the north, for example, have a different political agenda from that of non-Muslim nations in the interior.

Anwar el-Sadat, Nasser's successor, in time steered Egypt to friendly relations with the west. Few observers—undoubtedly, not even Sadat himself—would have predicted this eventual change in direction when he first took office. As an army officer during World War II, Sadat had been an anti-British activist, joining a plot with the Germans to oust the British from Egypt. Sadat had been arrested and imprisoned, but escaped. After the war, he became active in the Free Officers movement led by Nasser and Naguib.

When Sadat took office as president in 1970, he, like Nasser, at first tried to forge a union of Arab nations. Like Nasser, he failed. Then Sadat turned his attention inward, pressing for renewed social reforms. Meanwhile, like his predecessor, he voiced a militant attitude toward Israel and promised his people that Egypt would reclaim disputed territory.

Tension between the two countries flared into war again in October 1973. Egyptian forces attacked Israelis who still occupied the Sinai peninsula. Syria and other Arab countries joined Egypt's offensive. Fighting was bloody on both banks of the Suez Canal. This is remembered as the Yom Kippur War because Egypt launched its surprise attack on that annual Jewish holy day, also known as the Day of Atonement.

This time the United States and Soviet Union, fearful of an expanded conflict, worked through the United Nations to arrange a peace agreement. A buffer zone between the countries was patrolled by United Nations troops. With United States help, Egypt eventually negotiated the withdrawal of Israeli forces from key disputed areas. It was the first time in the history of the Arab-Israeli conflicts that an Arab country came away with a slice of reclaimed territory. This achievement momentarily raised Egypt's esteem greatly in the eyes of the Arab world.

Egypt had adopted a new constitution in September 1971. It described a socialist form of government and specified Islam as the country's official religion, Arabic as the national language. The constitution ensured equal rights for all citizens, freedom of public assembly, and the right to organize unions and other groups. Egyptian citizens 18 and older were given the right to vote; in fact, adult males were *required* to vote.

At the time, the Arab Socialist Union was the only political party approved by the Egyptian government; other parties had been banned in 1961. Not until 1977 were opposition parties permitted to function freely. By the 1990s the National Democratic Party had gained control. Among the major modern-day opposition groups are the New Wafd Party, the Muslim Brotherhood, and the Socialist Labor Party.

The government still keeps close rein on political activity. For example, parties are not allowed to organize based on class distinction. Among the results of such restrictions are somewhat less political fragmentation in Egypt than in other emerging countries and, arguably, a more stable government.

Western nations for the past half century have been leery of Egypt. In its first years of independence, the republic relied heavily on Soviet support—specifically, military equipment and advisers. It quickly developed bitter relations with England and France in the 1956 Suez crisis.

In the early 1970s, however, Egypt began taking steps to reduce Soviet influence. Although Egypt used Soviet arms and advisers during the 1973 conflict, Sadat the previous year had ordered 20,000 Russian military personnel in his country to return home. After that war, Sadat exchanged formal visits with the American president. He began seeking foreign investments in Egypt, a move welcomed by western companies. Additionally, America began pouring millions of dollars of foreign aid into Egypt.

Sadat made a far bolder move in 1977: he went to Israel to urge a permanent peace agreement between the two countries. Such an appeal would have been unheard of in the days of Nasser—and it cost Egypt the respect and friendship of many Arab governments. Furious at Sadat's initiative, they conspired to expel Egypt from the Arab League.

Undaunted, Sadat was convinced peaceful coexistence with Israel was essential to Egypt's well-being. He came to the United States in 1978 and 1979 to meet with Israeli Prime Minister Menachim Begin. After emotional negotiations, they signed the historic Camp David Accords, which called for the eventual withdrawal of Israeli forces from the Sinai Desert. Other Arab nations disapproved of the agreements, however. Radical Muslims called Sadat a trader. Tragically, his diplomatic policies soon cost him his life. He was assassinated in 1981 by Muslim extremists in the Egyptian army.

Vice President Hosni Mubarak, a former Egyptian air force general, succeeded Sadat and basically continued his policies. Mubarak has sought a permanent peaceful solution to the "Sinai Question." During his 20 years in office, though, Egypt has renewed friendly relations with most Arab countries. Egypt was admitted to the Islamic Conference in 1984. Subsequently, tension has resumed between Egypt and Israel.

During the Sadat administration, United States–Egyptian relations grew relatively strong—a reversal of the country's pro-Soviet stance of the 1950s and 1960s. Under Mubarak, Egypt has been cautious in its stance toward Israel and willing to cooperate with the west in important matters. While a staunch leader among the Arab nations, Egypt maintains a certain independence even within the explosive realm of Middle Eastern religious issues. In the 1990–1991 Persian Gulf conflict, for example, after Iraq invaded neighboring Kuwait, millions of Muslims denounced United States intervention. Egypt, on the other hand, was one of the Arab nations that sent troops into the region not to support defiant Iraq, but to oppose it.

Arguably, money had something to do with Egypt's alliance to the American-led anti-Iraqi coalition. After Kuwait successfully was restored, much of Egypt's $20 billion foreign debt was canceled. Still, the episode reflected not only Egypt's shrewdness but its determination to stand for justice in international affairs, even in the face of dangerous risks. And the country did pay a price. Muslim fundamentalists began carrying out bloody assaults against Egyptian government officials and Christians—even against tourists. Tourist revenues plummeted, prompting the government to break up some of the violent organizations beginning in 1993 and executing many of the extremist leaders.

Unrest continues in Egypt, as it does in other parts of the Arab world. Dissidents tried to assassinate Mubarak in 1995. More than 60 tourists were slaughtered in the city of Luxor in 1997. Egypt is free from foreign dominion . . . but not from internal threats of anarchy.

Sphinx, c. 1870 (Hippolyte Arnous) *The colossal sphinx at Giza is an important image in Egyptian art and legend. It dates back to the third century B.C. In ancient Egyptian religion, gods were depicted in a combination of human and animal form—that is, with the head of an animal on a human body. The opposite, a human head with a lion's body, was reserved for kings.*

7

Egypt Today

Probably no African nation was more eager for independence, or more ready, than Egypt. No intensive study of its past, no shrewd or mystical visionary, can prepare any fledgling nation for the trials of self-government. Each new day writes an unforeseen paragraph. But Egypt learned more quickly than others to stand on its own. It had to, as the keeper of one of the world's most important canals and as the connecting point between two continents.

What is life in Egypt like today? Let's take a look.

The Lay of the Land

Egypt is a dry land with little rainfall. Apart from certain mountain areas, especially in the southwestern corner and along the Red Sea in the east, it is largely desert dunes. This is the eastern section of Africa's great Sahara Desert. Were it not for the Nile, few inhabitants would be found in Egypt's interior.

The Nile flows out of the hills of Uganda and Ethiopia, some 4,000 miles from the Mediterranean Sea. In Egypt, it is lined by cliffs between the Aswan High Dam and Cairo. The lower 100 miles of it, from just north of Cairo to the sea, flattens into a fertile delta. Here the river splits into many streams that spread across an area up to 160 miles wide at the coast. The main river channels through the delta are the Rosetta Branch and the Damietta Branch.

Although the country beyond the Nile is desert on both sides, it is not entirely inhospitable. Bedouins and other tribes live and wander in the Egyptian deserts. Occasionally, we come to a curiously green, inviting reprieve—an island of life amid the scorching dunes. This is an oasis. Here, water springs from below the ground to form pools and to nourish beautiful trees and flowers. Small villages thrive at these places. Tribes bring their herds of camels, sheep, and goats here for sustenance.

Oasis water is said to be especially rich in mineral content. It is believed to provide rare health benefits for people with muscular, digestive, and other physical problems. People visit from foreign countries to bathe in these desert pools.

RELIGION

We've learned that the coming of the Arabs in the seventh century A.D. brought the Muslims. The Muslim faith remains by far the dominant religion in Egypt to this day. Egypt is part of the Arab nation, and Islam is the main form of Muslim worship.

Islam is, in fact, Egypt's state-sanctioned religion. Some 90 percent of Egyptians are Muslims, making Egypt one of the world's foremost Muslim countries. In Cairo, Al-Azhar University, more than a thousand years old, has become a major center for Islamic studies. Incidentally, it is the world's oldest continuous institution of higher learning.

Egypt's government exerts notable control over Islam. The government appoints upper-level Muslim leaders. Islam, in turn, pervades Egyptian life. Islam requires its followers to pray several times each day, to fast, and to give alms to those in poverty. Many Egyptians dress according to Islamic tradition: women commonly are seen with veils over their hair and shoulders; some men wear skullcaps and grow beards. Those who can afford it make periodic pilgrimages to Mecca, the Islamic center in Saudi Arabia. Egyptian mosques are used both for worship and social gatherings.

Bedouin, c. 1870 (G. Lekegian)

The Muslim faith is, in a sense, the "glue" that binds most of Egypt's diverse peoples. It dominates both the modernized urban areas and the ancient ways of many of the villagers.

Coptic Christianity survived after the Arab invasion, however. Today, Egypt's population includes several million Copts. Incidents of persecution against the Copts by antagonistic Muslims sometimes occur, even though Egyptian law guarantees the right of religious minorities to worship.

Water Carrier, c. 1870 (G. Lekegian) *A barefoot Egyptian carrier mounts stone steps with a water-filled animal skin on his back. What makes this picture remarkable is both its age and the way in which the photographer clearly understood and exploited the artistic potential of the new medium.*

GOVERNMENT

Egypt is ruled by a president and a national legislature called the People's Assembly. The assembly consists of 454 representatives, 444 of them elected and 10 selected by the president.

The Egyptian president is elected, but not the same way as in America. The assembly nominates an individual for the position, and a public vote is held to accept or reject the nomination. The president serves a six-year term and can be renominated for subsequent terms. Egypt is unique among African nations—actually, among most nations of the world—in that it has had only three presidents during its entire half century of independence (not counting the figurehead president of the early 1950s, General Naguib).

Assembly representatives serve five-year terms in office. Interestingly, the nation's constitution requires that at least half the assembly members come from the working class. It may seem that with its power to nominate the president, the People's Assembly is the dominant component of Egypt's government. In reality, the Egyptian president wields more power than the heads of state in many other republics. Egypt's president appoints the government's highest officials, who in turn appoint lower administrators. Thus, the president controls appointments throughout the nation's governing structure. Most government leaders belong to the National Democratic Party, the nation's largest political organization.

Egypt has another government body called the Majlis ash-Shura, which serves only to advise the government leaders. This group consists of 258 members, some appointed and some elected.

To help administer the government, the president names a vice-president (possibly more than one) and a council of ministers. A prime minister is chosen to lead the council.

At the regional level, Egypt is divided into twenty-six "governorates." The president appoints each governor—a further extension of the president's influence throughout the country. Egypt's court system is somewhat similar to America's, except that there are no jury trials. The ultimate body of appeal in legal matters is the Supreme Constitutional Court. Below it, the judicial structure includes levels of regional and district courts, as well as lower courts of appeal. A cabinet member, the minister of justice, suggests candidates for judgeships and the president appoints them. This is much like the appointive process in the United States, where the president relies on the United States attorney general to help select federal judges.

Water Carrier, Cairo, c. 1870 (G. Lekegian)

The Egyptian government oversees a standing army, air force, and navy. Egypt uses the draft to maintain personnel levels within its military. Draftees are required to serve for three years. Altogether, some 440,000 Egyptians staff the armed forces. This is a comparatively large military institution for a nation of Egypt's population. The government deems it neces-

Arab Woman and Child, c. 1870 (G. Lekegian)

sary because of continual uneasiness between the Arab countries and Israel.

MAJOR CITIES

Egypt's capital, Cairo, is not only the country's primary metropolis; it is the largest city on the African continent. More than 15

Egyptian young Women, c. 1870 (Hippolyte Arnous)

million people live in the greater Cairo area. Cairo dates to the tenth century A.D.

Giza, on the west bank of the Nile near Cairo, is itself a large city. It also is a primary tourist destination, for it is here that we find the Great Sphinx as well as several pyramids. The ancient Sphinx is a gigantic stone monument depicting a pharaoh's head atop a lion's body.

Alexander the Great, the young Macedonian emperor who conquered much of the eastern Mediterranean and Middle East in the fourth century B.C., at one time had no fewer than thirty cities named after him. The greatest surviving one is the Egyptian port city in the Nile delta. Alexandria is Egypt's second-largest city.

In ages past, Alexandria was known for its great stone lighthouse (one of the seven wonders of the ancient world) and its massive library, which contained thousands of handwritten scrolls and books. Timeless examples of the city's glorious architecture can be seen today—and many gigantic stone statues and columns recently were found by divers in the waters of the bay. Between Alexandria and Cairo are the ruins of Sais, capital of Egypt during the twenty-sixth dynasty.

Just as Americans like to visit the beach in summer, so do Egyptians. Well-to-do citizens who live in cities up the Nile retreat to the Alexandria area, where the climate is slightly cooler and summer winds off the Mediterranean provide relief from seasonal misery.

Egypt's other major seaports are Port Said and Suez, at either end of the Suez Canal. Port Said, at the Mediterranean end, owes its existence to the canal. It was created when workers began building the canal almost a century and a half ago. Today it stands as an important world port.

Where the People Live

In modern Egypt, the population has been shifting toward the cities. People rooted for generations in remote village life now hope to find jobs in urban areas and improve their standard of living. Overall, meanwhile, the country's population also has grown tremendously in the last 50 years. As a result of these trends, overcrowding has become a problem in Egypt's cities.

All but a tiny fraction of Egyptians live along the River Nile and the Suez Canal. Those regions, in fact, have become two of the heaviest-populated areas in the world.

People from Sais, c. 1870 (G. Lekegian) *The ruins of Sais are located between Alexandria and Cairo. The city was the capital of Egypt during the twenty-sixth dynasty (664–525 B.C.). When Herodotus visited Sais in the fifth century, it still was considered one of the finest cities in Egypt. However, inscribed stones found on the site are all that remain of ancient Sais. In 1906 Baedecker advised its readers, "A visit . . . can hardly be recommended even for the specialist."*

Agriculture

Despite the general population shift into large towns and cities, farming remains the nation's leading occupation. More than a third of Egypt's workers are farmers, even though farming generates only about a sixth of the nation's gross domestic product.

Practically all Egyptian farms are found near the River Nile. Most are owned by citizens, but the government is closely involved in farming policies. For example, it requires farmers to produce a certain percentage of their vegetables, grains, and fruits for Egypt's internal use, rather than sell it all to exporters.

Egyptians produce high volumes of cotton, rice, fruits, and sugar cane. One unusual crop, of which Egypt is the world's top producer, is dates. The interesting thing about dates is that they're among the few marketable crops that thrive in desert oases.

Sheep and goats are the main types of livestock raised by Egyptian farmers. They yield three forms of saleable goods: meat, wool, and milk. Other common beasts of burden include buffalo, cattle, and donkeys. Poultry is another important Egyptian farm product.

Industry

Oil is one of Egypt's most important resources, as it is in many other arid countries. Egypt drills some of its oil from the deserts, but most from its offshore waters in the Red Sea and Gulf of Suez. Oil is one of the nation's three main energy sources, along with natural gas and hydroelectric power from the Aswan High Dam. Petroleum also is a leading export item, bringing the country substantial revenues.

Factories, most of which are controlled by the government, employ about 20 percent of Egypt's workers. Food processing, textiles, and steel are the major manufacturing operations. Other factories produce fertilizer, chemicals, and medicine.

Tourism is a thriving industry, despite the fact that Egypt is mostly desert. Egyptians did not have to create their tourist attractions; they have existed for centuries: the pyramids and other age-old monuments, as well as fascinating architecture, homes, and mosques.

Coppersmith, Cairo, c. 1870 (G. Lekegian)

THE ECONOMY

Besides oil, other major Egyptian exports include cotton and fruit. However, the country must import more, dollar for dollar, than it exports. Imports include certain food types and machinery. When a country must buy more than it can produce, foreign debt results. One of the most serious problems facing Egypt today is the same as the crisis that gave the British increased control over the country in the 1870s: mounting national debt. Egypt's leading trading partners include the United States, Germany, Italy, and France.

Apart from petroleum and natural gas, Egypt is not particularly rich in natural resources. It mines iron ore and a few other minerals, but mining is not a major part of the country's economy.

Almost half the jobs in Egypt are in what's known as "service industries." These are staffed by people who perform needed services—for example, in government, schools, the transportation and communication systems, banking institutions—but do not produce marketable commodities.

Transportation

From the beginning of Egyptian civilization, the River Nile has provided the principal means of transportation through the country. While plane, train, bus, and automobile traffic has broadened and improved modern-day travel, boating commerce along the Nile remains important. A system of canals provides additional commercial links by boat.

Highways and rail lines through the desert connect the country's cities. Citizens often travel between cities by bus and train. Only one in approximately fifty Egyptians owns a car. We see bicycles and even donkey carts joining motor vehicles on the highways.

A major international airport is at Cairo. The national airline, EgyptAir, is operated by the government. At the opposite extreme of travel modes, camel caravans still ply the desert dunes as they have for thousands of years.

Schools and Colleges

Egypt's education system is more advanced than in certain other African countries, but it needs much improvement. Only about 50 percent of Egyptian adults can read and write. In some rural areas, the illiteracy rate is especially high. The government requires students aged six to fourteen to attend grade school. Many schools are overcrowded and have too few teachers. New schools need to be built, especially in the smaller villages, but funds are lacking.

Of those students who finish grade school, about half continue to secondary or high school, or to intermediate-level tech-

Arab Cafe, Cairo, c. 1870 (G. Lekegian)

nical school. One in five go to college. Cairo University is the largest of the nation's thirteen universities.

MEDIA AND ENTERTAINMENT

We find one television set for every twelve people in Egypt. By American standards, that is a low ratio; Americans are used to TV sets everywhere they turn, at home and in public. But compared to other African nations, it is a fairly high number. Egypt has one radio for every three people.

Cairo is virtually the publishing capital of the Arab world and is home to the Middle East News Agency. About fifteen newspapers are published in Egyptian cities. They are not completely controlled by the government, but criticism of those in power is not as common as in American newspapers. The Egyptian government owns the nation's broadcast media.

Artistically, Egypt's contributions to society date from the dynastic period. We have only to gaze at the enormous monu-

Carpenter's Shop, Cairo, c. 1870 (G. Lekegian)

ments at Giza or elsewhere, or view pictures of the carvings and paintings inside the ancient tombs and later-day mosques, to appreciate Egyptian artistry. And Egyptians continue to produce art, music, and literature of international significance. The country supports national theatre, dance, and symphony companies. Much of the best Arabic filmmaking and writing comes from Egypt. One of the country's leading authors, Naguib Mahfouz, in 1988 won the Nobel Prize—the first time any work written in Arabic had earned this distinction.

Musically, Egyptian composers and performers plainly have been influenced by globally popular western sounds. However, they like to merge eastern and western elements for a unique

Cafe At Giza, c. 1870 (J. Pascal Sebah) *Baedeker's* Egypt *recommended at least half a day to see the pyramids at Giza with "sun umbrellas and smoked spectacles" urged as "precautions against the glare of the sun." The 180-room Mena House Hotel and its restaurant—both still in existence—received an outstanding review, but otherwise travelers were strongly advised to bring their own food.*

musical style. They also have preserved "pure" music from their country's past.

LIFESTYLES

In many ways, life is quite different in Egyptian cities than it is in remote villages. The jobs, dress, and daily routines of many city dwellers are not altogether unlike those observed in western industrialized nations—aside from the visible aspects of the

Muslim influence (regulated prayers, strict dress standards among the more orthodox Muslims, etc.). Likewise, heavy traffic, overcrowded streetcars and living quarters, crime, drugs, and other problems challenge the cities, much as they do in western cities.

Cairo and other Egyptian cities have a small contingent of wealthy citizens but are marred by appalling slums. Flimsy hovels—some perched on rooftops, some set on other people's property—are the homes of many of the poor. The homes of the wealthy provide a startling contrast of opulence, just as they did a century or two ago. Outside Cairo, ancient tombs provide shelter to a small city of indigents.

In rural areas, on the other hand, you glimpse life that remains largely the same as in centuries past. The people work farms and herd livestock. Men, women, and children all join in the daily tasks. These people are the modern-day fellahin, the peasant working class whose toils for centuries have produced the wealth and monuments of the Nile. Some own their small farms; the poorer ones work for others. They live in small, plainly furnished huts of stone or mud with straw roofs. Just as in other countries, where people in small towns tend to "know everyone," Egypt's rural villages evidence close-knit families and neighbors.

As in most African countries, socializing is the favorite pastime of Egyptians. They often meet in city and village marketplaces called bazaars, not just to shop but to talk. They also like to gather at homes over cups of tea or coffee. Wealthy Egyptian families enjoy varied diets, including imported items and different meat types. Common people, on the other hand, dine simply. A village family may sit around a bowl of stew and dip bread into it. Many meals include a bean preparation called *ful.*

Egyptians of different backgrounds and settings wear a variety of clothing styles. In the cities, you will see some of the natives—typically, those who are well-to-do—dressed much the same as in America, alongside poorer people and orthodox Muslims in simple, flowing garments and traditional Arab

Arab Dancer, c. 1870 (G. Lekegian)

costumes. Some choose gaily colored garments; others wear
dark, conservative attire.

WILDLIFE

Since the completion of the Aswan High Dam in 1971, Lake
Nasser has been teeming with bird life: geese, plovers, pelicans,
terns, and other species. Crocodiles—timeless dwellers of the
Nile—idle in shallow water. Herons, ospreys, spoonbills, and
other birds are seen along the river and the edge of the Red Sea.
Egypt is home to some 300 species of birds.

Camels, called "ships of the desert" because they can live for
days and weeks without drinking water, are common not only
in the deserts but in cities and towns. For ages, camels were the

Arab Singers, c. 1870 (G. Lekegian)

primary means of travel through the arid Middle East. When they do drink water, an adult camel can consume as much as twenty-five gallons at a time.

Ibex, similar to mountain goats, and other animals inhabit the Sinai peninsula. Among the insect realm, the dung beetle, or scarab, is held in high regard by many Egyptians. Famous for rolling balls of manure several times its size across the sands, the scarab was thought by the ancients to represent the god of the dawn, Khepri.

Also once held sacred is the cat, which first was domesticated in Egypt. Feral cats now roam the Nile Delta. Ancient Egyptians believed the cat represented Bast, daughter of Ra, the sun god. They prepared mummies of dead cats.

Besides the notorious Nile crocodiles, Egypt's reptiles include deadly snakes—the horned viper and asp. The country has about 100 species of fish.

The date palm for ages has been Egypt's most prevalent native tree. Tamarisk, carob, and sycamore also are common in the watered fringes of the country, along with varieties imported long ago from other regions such as the eucalyptus, cypress, and mimosa. Wildflowers and grapes are bountiful in the Nile delta.

Interior of a Wealthy Arab House, c. 1870 (G. Lekegian)

A STEADFAST ANCHOR IN A SEA OF PROBLEMS

Many of Egypt's problems today stem from its large population, growing at a rate of more than two percent each year. The unemployment rate also is high. Housing is inadequate in many places. These conditions result in poverty and hunger among many of the people.

Religion is another cause of unrest. Most Muslims in Egypt follow the ancient, strict Sunni form of Islamic worship. Some devout Muslims believe their government and society have been corrupted by western influences and have strayed from the true faith. Their protests at times have turned violent.

The country's economy is not as stable as its leaders would like. As in other oil-producing countries, Egypt's economy improves or worsens with changing world oil prices. A price drop in 1986, for example, aggravated the country's already severe national debt and increased its reliance on foreign aid.

Despite their problems, though, Egyptians have some cause for optimism. Progress has been made in health care, for example, with expanded treatment facilities both in the cities and the rural areas. Age-old diseases that still plague certain other African countries—smallpox, cholera, malaria—have been all but conquered in Egypt.

Egyptians have one source of constant assurance available to few other nations: their incredible past. Their ancestors endured torturous treatment by stern pharaohs and survived century after century of foreign domination. In the shadow of such a history, all the problems of Egypt's recent independence seem perhaps not so great after all.

CHRONOLGY

639 A.D.	Arabs bring Egypt under Muslim control.
1250	Mamelukes—freed Turkish slaves—take power in Egypt.
1798	The British navy defeats the French navy at the Battle of the Nile.
1805	Muhammad Ali becomes pasha in Egypt.
1863	Ismail Pasha, Muhammad Ali's nephew, becomes viceroy.
1869	Completion of the Suez Canal after a decade of construction.
1875	Ismail Pasha sells Egyptian shares in the Suez Canal to Great Britain in order to resolve some of Egypt's debt.
1882	Lethal rioting in Alexandria by militant nationalists; England responds by establishing military force.
1899	The Egyptian/Sudanese "condominium" arrangement is established, under which both Great Britain and Turkey jointly exerted certain controls.
1906	The Dinshwai incident—a peasant uprising leads to severe British reprisals.
1914	World War I ends British/Turkish cooperation. Britain declares Egypt a British "protectorate."
1918	The Wafd political party begins a growing movement for native control over Egyptian affairs.
1922	Though still heavily controlled by Great Britain, Egypt becomes something of an independent "country" with its own king, Fuad I.
1937	King Farouk ascends to the throne; Great Britain gives up its control over international citizens in Egypt.
1948–49	Israel becomes the arch-enemy of Egypt and other Muslim nations.
1952	Opposition to British influence results in the infamous Cairo riots. The Free Officers, led by Gamal Abdel Nasser, ousts King Farouk and takes control of Egypt.
1956	The Suez Canal crisis causes superpower tension and temporarily closes the great canal.
1967	The Six-Day war between Egypt and Israel.
1970	Nasser dies; Vice-President Anwar el-Sadat becomes Egypt's leader.
1971	Completion of the Aswan High Dam on the River Nile.
1973	Egypt and other Arab countries initiate the Yom Kippur War against Israel.

CHRONOLGY

1978–79	Meetings in the U.S. between Sadat and Israeli leader Menachim Begin result in the Camp David Accord between their two countries.
1981	Sadat is assassinated; Vice-President Hosni Mubarak becomes president; Mubarak's administration continues today.
1990–91	Egypt joins U.S.-organized forces against Iraq in the Persian Gulf War.
1995	Assassins are unsuccessful in killing President Mubarak.
1997	Dissidents slaughter 60 tourists in Luxor.

GLOSSARY

anarchy—a society with no system of authority or order

archaeologist—one who preserves and studies past human remains

aristocrat—a member of a country's upper or ruling class

bureaucrat—a government bureau official

caravan—a convoy of travelers or transport animals/vehicles; camel caravans still are useful in parts of Africa

coalition—a usually short-term union of political factions

commodity—an item of trade

consul—a government's chief representative in a foreign colony

coup—a political take-over, which may be violent or peaceful

delta—the fertile, low-lying area where a river fans out, often forming many channels, near its mouth

demolition—destroying an obstacle or building with explosives

demotic—a common, simplified form of ancient writing similar to **hieroglyphics**

dialect—a variation of a language spoken in a specific region

dynasty—a blood-related line of rulers

fellahin—peasant laborers

Gatling gun—an early type of rapid-firing machine gun

gross national product—the total value of what a country produces in a certain time span

hydroelectric power—electrical power generated by fast-flowing water

hieroglyphics—an ancient form of writing that used pictures and symbols

industrialized nations—countries—notably western nations—whose economies are based on highly developed industries

insurgent—rebel

isthmus—a thin strip of territory lying between two major land masses

mosque—place of worship for Muslims

mufti—a country's highest Muslim legal scholar

nomad—a wanderer; a member of a tribe or group who move from one area to another with seasonal changes, herding or hunting

oasis—an "island of life" in a desert, with water and plants

parliament—a body of governing representatives

persecution—unfair and often violent—even fatal—treatment of one group of people by another

pharaoh—ancient Egyptian ruler

protectorate—a country or region under the control and "protection" of a foreign power

pyramid—one of the huge stone monuments in Egypt, dating to ancient times, angling upward to a point from a vast rectangular base

GLOSSARY

republic—a country ruled not by a king or queen, but by a "popular" government (military factions and dictators have seized power in many African republics)

silt—a fine sediment consisting of particles tinier than sand

socialist—a type of society and government in which the people jointly own business, farming, and industrial interests

sultan—a Muslim ruler

toll—a fee, or tariff, charged by a government or governing agency for passage over a bridge or section of highway, or through a waterway

treaty—a formal trading and/or peace agreement between nations or, as in Africa during colonization, between European powers or trading companies and important native chiefs

usurious interest—a blatantly high, sometimes illegal, rate of interest

viceroy—a governor of a colony or province, appointed by the foreign government that controls the region

WORLD WITHOUT END

DEIRDRE SHIELDS

ONE SUMMER'S DAY in 1830, a group of Englishmen met in London and decided to start a learned society to promote "that most important and entertaining branch of knowledge—Geography," and the Royal Geographical Society (RGS) was born.

The society was formed by the Raleigh Travellers' Club, an exclusive dining club, whose members met over exotic meals to swap tales of their travels. Members included Lord Broughton, who had travelled with the poet Byron, and John Barrow, who had worked in the iron foundries of Liverpool before becoming a force in the British Admiralty.

From the start, the Royal Geographical Society led the world in exploration, acting as patron and inspiration for the great expeditions to Africa, the Poles, and the Northwest Passage, that elusive sea connection between the Atlantic and Pacific. In the scramble to map the world, the society embodied the spirit of the age: that English exploration was a form of benign conquest.

The society's gold medal awards for feats of exploration read like a Who's Who of famous explorers, among them David Livingstone, for his 1855 explorations in Africa; the American explorer Robert Peary, for his 1898 discovery of the "northern termination of the Greenland ice"; Captain Robert Scott, the first Englishman to reach the South Pole, in 1912; and on and on.

Today the society's headquarters, housed in a red-brick Victorian lodge in South Kensington, still has the effect of a gentleman's club, with courteous staff, polished wood floors, and fine paintings.

AFTERWORD

The building archives the world's most important collection of private exploration papers, maps, documents, and artefacts. Among the RGS's treasures are the hats Livingstone and Henry Morton Stanley wore at their famous meeting ("Dr. Livingstone, I presume?") at Ujiji in 1871, and the chair the dying Livingstone was carried on during his final days in Zambia. The collection also includes models of expedition ships, paintings, dug-out canoes, polar equipment, and Charles Darwin's pocket sextant.

The library's 500,000 images cover the great moments of exploration. Here is Edmund Hillary's shot of Sherpa Tenzing standing on Everest. Here is Captain Lawrence Oates, who deliberately walked out of his tent in a blizzard to his death because his illness threatened to delay Captain Scott's party. Here, too is the American Museum of Natural History's 1920 expedition across the Gobi Desert in dusty convoy (the first to drive motorised vehicles across a desert).

The day I visited, curator Francis Herbert was trying to find maps for five different groups of adventurers at the same time from the largest private map collection in the world. Among the 900,000 items are maps dating to 1482 and ones showing the geology of the moon and thickness of ice in Antarctica, star atlases, and "secret" topographic maps from the former Soviet Union.

The mountaineer John Hunt pitched a type of base camp in a room at the RGS when he organised the 1953 Everest expedition that put Hillary and Tenzing on top of the world. "The society was my base, and source of my encouragement," said the late Lord Hunt, who noted that the nature of that work is different today from what it was when he was the society's president from 1976 to 1980. "When I was involved, there was still a lot of genuine territorial exploration to be done. Now, virtually every important corner—of the land surface, at any rate—has been discovered, and exploration has become more a matter of detail, filling in the big picture."

The RGS has shifted from filling in blanks on maps to providing a lead for the new kind of exploration, under the banner of geography: "I see exploration not so much as a question of 'what' and 'where' anymore, but 'why' and 'how': How does the earth work, the environment function, and how do we manage our resources sustainably?" says the society's director, Dr. Rita Gardner. "Our role today is to answer such questions at the senior level of scientific research," Gardner continues, "through our big, multidisciplinary expeditions, through the smaller expeditions we support and encourage, and by advancing the subject of geography, advising governments, and encouraging wider public understanding. Geography is the subject of the 21st century because it embraces everything—peoples, cultures, landscapes, environments—and pulls them all together."

The society occupies a unique position in world-class exploration. To be invited to speak at the RGS is still regarded as an accolade, the ultimate seal of approval of Swan, who in 1989 became the first person to walk to both the North and South Poles, and who says, "The hairs still stand on the back of my neck when I think about the first time I spoke at the RGS. It was the greatest honour."

The RGS set Swan on the path of his career as an explorer, assisting him with a 1979 expedition retracing Scott's journey to the South Pole. "I was a Mr. Nobody, trying to raise seven million dollars, and getting nowhere," says Swan. "The RGS didn't tell me I was mad—they gave me access to Scott's private papers. From those, I found fifty sponsors who had supported Scott, and persuaded them to fund me. On the basis of a photograph I found of one of his chaps sitting on a box of 'Shell Spirit,' I got Shell to sponsor the fuel for my ship."

The name "Royal Geographical Society" continues to open doors. Although the society's actual membership—some 12,600 "fellows," as they are called—is small, the organisation offers an incomparable net-work of people, experience, and expertise. This is seen in the work of the Expeditionary Advisory Centre. The EAC was established in 1980 to provide a focus for would-be explorers. If you want to know how to raise sponsorship, handle snakes safely, or find a mechanic for your trip across the Sahara, the EAC can help. Based in Lord Hunt's old Everest office, the EAC funds some 50 small expeditions a year and offers practical training and advice to hundreds more. Its safety tips range from the pragmatic—"In subzero temperatures, metal spectacle frames can cause frostbite (as can earrings and nose-rings)"—to the unnerving—"Remember: A decapitated snake head can still bite."

The EAC is unique, since it is the only centre in the world that helps small-team, low-budget expeditions, thus keeping the amateur—in the best sense of the word—tradition of exploration alive.

AFTERWORD

"The U.K. still sends out more small expeditions per capita than any other country," says Dr. John Hemming, director of the RGS from 1975 to 1996. During his tenure, Hemming witnessed the growth in exploration-travel. "In the 1960s we'd be dealing with 30 to 40 expeditions a year. By 1997 it was 120, but the quality hadn't gone down—it had gone up. It's a boom time for exploration, and the RGS is right at the heart of it."

While the EAC helps adventure-travellers, it concentrates its funding on scientific field research projects, mostly at the university level. Current projects range from studying the effect of the pet trade on Madagscar's chameleons, to mapping uncharted terrain in the south Ecuadorian cloud forest. Jen Hurst is a typical "graduate" of the EAC. With two fellow Oxford students, she received EAC technical training, support, and a $2,000 grant to do biological surveys in the Kyabobo Range, a new national park in Ghana.

"The RGS's criteria for funding are very strict," says Hurst. "They put you through a real grilling, once you've made your application. They're very tough on safety, and very keen on working alongside people from the host country. The first thing they wanted to be sure of was whether we would involve local students. They're the leaders of good practice in the research field."

When Hurst and her colleagues returned from Ghana in 1994, they presented a case study of their work at an EAC seminar. Their talk prompted a $15,000 award from the BP oil company for them to set up a registered charity, the Kyabobo Conservation Project, to ensure that work in the park continues, and that followup ideas for community-based conservation, social, and education projects are developed. "It's been a great experience, and crucial to the careers we hope to make in environmental work," says Hurst. "And it all started through the RGS."

The RGS is rich in prestige but it is not particularly wealthy in financial terms. Compared to the National Geographic Society in the U.S., the RGS is a pauper. However, bolstered by sponsorship from such companies as British Airways and Discovery Channel Europe, the RGS remains one of Britain's largest organisers of geographical field research overseas.

The ten major projects the society has undertaken over the last 20 or so years have spanned the world, from Pakistan and Oman to Brunei and Australia. The scope is large—hundreds of people are currently working in the field and the emphasis is multidisciplinary, with the aim to break down traditional barriers, not only among the different strands of science but also among nations. This is exploration as The Big Picture, preparing blueprints for governments around the globe to work on. For example, the 1977 Mulu (Sarawak) expedition to Borneo was credited with kick-starting the international concern for tropical rain forests.

The society's three current projects include water and soil erosion studies in Nepal, sustainable land use in Jordan, and a study of the Mascarene Plateau in the western Indian Ocean, to develop ideas on how best to conserve ocean resources in the future.

Projects adhere to a strict code of procedure. "The society works only at the invitation of host governments and in close co-operation with local people," explains Winser. "The findings are published in the host countries first, so they can get the benefit. Ours are long-term projects, looking at processes and trends, adding to the sum of existing knowledge, which is what exploration is about."

Exploration has never been more fashionable in England. More people are travelling adventurously on their own account, and the RGS's increasingly younger membership (the average age has dropped in the last 20 years from over 45 to the early 30s) is exploration-literate and able to make the fine distinctions between adventure / extreme / expedition / scientific travel.

Rebecca Stephens, who in 1993 became the first British woman to summit Everest, says she "pops along on Monday evenings to listen to the lectures." These occasions are sociable, informal affairs, where people find themselves talking to such luminaries as explorer Sir Wilfred Thesiger, who attended Haile Selassie's coronation in Ethiopia in 1930, or David Puttnam, who produced the film *Chariots of Fire* and is a vice president of the RGS. Shortly before his death, Lord Hunt was spotted in deep conversation with the singer George Michael.

Summing up the society's enduring appeal, Shane Winser says, "The Royal Geographical Society is synonymous with exploration, which is seen as something brave and exciting. In a sometimes dull, depressing world, the Royal Geographical Society offers a spirit of adventure people are always attracted to."

ABOUT THE AUTHORS

Dr. Richard E. Leakey is a distinguished paleo-anthropologist and conservationist. He is chairman of the Wildlife Clubs of Kenya Association and the Foundation for the Research into the Origins of Man. He presented the BBC-TV series *The Making of Mankind* (1981) and wrote the accompanying book. His other publications include *People of the Lake* (1979) and *One Life* (1984). Richard Leakey, along with his famous parents, Louis and Mary, was named by *Time* magazine as one of the greatest minds of the twentieth century.

Daniel E. Harmon is an editor and writer living in Spartanburg, South Carolina. The author of several books on history, he has contributed historical and cultural articles to *The New York Times, Music Journal, Nautilus,* and many other periodicals. He is the associate editor of *Sandlapper: The Magazine of South Carolina* and editor of *The Lawyer's PC* newsletter.

Deirdre Shields is the author of many articles dealing with contemporary life in Great Britain. Her essays have appeared in *The Times, The Daily Telegraph, Harpers & Queen,* and *The Field.*

FURTHER READING

Casson, Lionel, et al. *Ancient Egypt* ("Great Ages of Man" series). New York: Time-Life Books. 1965.

Gifford, Prosser, and William Roger Louis, editors. *France and Britain in Africa: Imperial Rivalry and Colonial Rule.* New Haven, CT: Yale University Press. 1971.

Hallett, Robin. *Africa Since 1875: A Modern History.* Ann Arbor, MI: The University of Michigan Press. 1974.

Harmon, Daniel E. *Exploration of Africa: The Emerging Nations: Sudan.* Philadelphia: Chelsea House Publishers. 2001.

Harper, Paul. *The Suez Crisis* ("Flashpoints" series). England: Wayland Ltd. 1986.

Hubbard, Gardiner G. "The Evolution of Commerce." *National Geographic,* March 26, 1892, Page 1.

Jarvis, H. Wood. *Pharaoh to Farouk.* New York: The MacMillan Company. 1955.

Kallen, Stuart A. *Egypt* ("Modern Nations of the World" series). San Diego: Lucent Books. 1999.

La Riche, William. *Alexandria: The Sunken City.* London: Weidenfeld & Nicholson. 1996.

Marsot, Afaf Lutfi Al-Sayyid. *A Short History of Modern Egypt.* Cambridge, UK: Cambridge University Press. 1985.

Metz, Helen Chapin, editor. *Egypt: A Country Study.* Washington, DC: Federal Research Division, Library of Congress. 1991.

Outhwaite, Leonard. *Unrolling the Map.* New York: Reynal & Hitchcock. 1935.

Packenham, Thomas. *The Scramble for Africa, 1876-1912.* New York: Random House. 1991.

Palmer, R.R., editor. *Rand McNally Atlas of World History.* New York: Rand McNally & Company. 1965.

Simpich, Frederick. "Along the Nile, Through Egypt and the Sudan." *National Geographic,* October 1922, Page 379.

Stewart, Desmond. *Cairo: 5500 Years.* New York: Thomas Y. Crowell Company. 1968.

Whipple, A.B.C. *The Seafarers: Fighting Sail.* Alexandria, VA: Time-Life Books. 1978.

Williams, Maynard Owen. "The Suez Canal: Short Cut to Empires." *National Geographic,* November 1935, Page 611.

Encyclopedias:

Encyclopedia Britannica, World Book, Hutchinson Educational Encyclopedia 2000, Microsoft Encarta, Webster's World Encyclopedia 2000

INDEX

962 Har

2/24